Islands of the Ottomon Empire

Islands of the Ottomon Empire

ANTONIS HADJIKYRIACOU

Editor

Markus Wiener Publishers
Princeton

Copyright © 2018 by the Department of Near Eastern Studies, Princeton University

Reprinted from *Princeton Papers: Interdisciplinary Journal of Middle Eastern Studies*, Volume XVIII

All rights reserved. No part of this book may be reproduced or transmitted in any form or by any means, whether electronic or mechanical—including photocopying or recording—or through any information storage or retrieval system, without permission of the copyright owners.

For information write to:
Markus Wiener Publishers
231 Nassau Street, Princeton, NJ 08542
www.markuswiener.com

Library of Congress Cataloging-in-Publication Data

Names: Hadjikyriacou, Antonis, editor of compilation.
Title: Islands of the Ottoman Empire / Antonis Hadjikyriacou, editor.
Description: Princeton : Markus Wiener Publishers, [2018]
Identifiers: LCCN 2018016058| ISBN 9781558766372 (hardcover : alk. paper) | ISBN 1558766372 (hardcover : alk. paper) | ISBN 9781558766389 (pbk. : alk. paper) | ISBN 1558766389 (pbk. : alk. paper)
Subjects: LCSH: Islands—Turkey—History. | Turkey—History—Ottoman Empire, 1288-1918. | Turkey—History, Naval.
Classification: LCC DR486 .I86 2018 | DDC 956/.01509142—dc23
LC record available at https://lccn.loc.gov/2018016058

Markus Wiener Publishers books are printed in the United States of America on acid-free paper, and meet the guidelines for permanence and durability of the Committee on Production Guidelines for Book Longevity of the Council on Library Resources.

Contents

ANTONIS HADJIKYRIACOU
Envisioning Insularity in the Ottoman World vii

PART I: CONCEPTUALIZING INSULARITY 1

SPYROS ASDRACHAS
Observations on Insularity in the Greek World 3

ELEFTHERIA ZEI
The Historiography of Aegean Insularity 37

**PART II: VIOLENCE AND LAW IN
TERRAQUEOUS SPACE** 59

MICHAEL TALBOT
Separating the Waters from the Sea: The Place of Islands
in Ottoman Maritime Territoriality during the Eighteenth Century 61

MURAT CEM MENGÜÇ
Maritime Warfare in the Aegean and Ionian Islandscapes:
Safai's History of the 1499 Lepanto Expedition 87

PART III: REGULATING ISLANDS 105

FATMA ŞİMŞEK
Blockading an Island: Collective Punishment, Islanders,
and the State in the "Largest" Island at the End of the
Nineteenth Century 107

KAHRAMAN ŞAKUL
The Ottoman Peloponnese before the Greek Revolution:
"A Republic of *Ayan*, *Hakim* and *Kocabaşı*" in
"the Sea of Humans and Valley of Castles" 121

Envisioning Insularity in the Ottoman World

ANTONIS HADJIKYRIACOU

"The Mediterranean islands are more numerous and above all more important than is generally supposed."[1] To start a volume on Ottoman insularities with a quote from Braudel may seem an easy, if not banal, choice. Yet, as with so many instances from the wide array of Braudelian quotes, there is more to this aphorism than the simplicity that catches the eye. Why would the fact that Mediterranean islands are more numerous than generally supposed be interesting? If islands are quantitatively underestimated, this is because they are often taken for granted. Braudel warns against the instinctive assumption of familiarity with islands. This leads us to the second, qualitative part of the statement: that islands are more important than people think they are. This is not to argue for the importance of islands; it is to point out that they are often underestimated. In other words, Braudel advices us to go beyond the ostensible when thinking about islands. This volume aims to do precisely that: to enquire into the meaning of insularity, i.e. the condition of being an island, in the Ottoman world. Most importantly, the starting point of the contributions that follow is that islands are neither obvious nor self-explanatory geographical categories.

The Ottoman Turkish word for island is *cezire*, a word that can also mean peninsula. In Arabic, the word also denotes northwest Mesopotamia and the Arabian Peninsula; it thus can be related to any body of land partially or totally surrounded by water. Its Arabic root, (c–z–r) means to cut off or slaughter, as well as to sink, flow away or ebb.[2] Similar connotations of liquid movement exist in Greek – the language predominantly spoken in Ottoman islands. *Nisi* comes from the Ancient Greek *nisos*, the root of which is the verb *neo*: to float.[3] If English is currently the international language of academia, it may also be worth exploring the etymology of the word used by the contributors of this volume and many of their sources. From the old English word "ieg", meaning sea, and "land" we see the interplay between two opposing elements yet again. Land and sea do not define, but also merge with each other. Drawing inspiration from these linguistic observations, we may infer that islands entail a dynamic relationship between land and sea. After all, the sea that apparently delineates the coastline surrounding an island is in constant movement with its ebbs and flows.

In conventional understandings of space, the relationship between land and sea is often seen as mutually exclusive. The term we employ for the purposes of this volume is insularity. The literal meaning of insularity is isolation. In this sense, the sea is understood as the boundary that defines the island and the cause of isolation. To return to Braudel, he was one of the first to offer a caveat against such an exclusivist conceptualization of the relationship between land and sea whereby insularity is equated with isolation:

> That the sea surrounds the islands and cuts them off from the rest of the world more effectively than any other environment is certainly true whenever they are really situated outside the normal sea routes. But when they are integrated into shipping routes, and for one reason or another (often external and quite gratuitous reasons) become one of the links in a chain, they are on the contrary actively involved in the dealings of the outside world, less cut off from them than some inaccessible mountain areas [...] "Isolation" is a relative phenomenon [...] simultaneously with this isolation and in striking contrast, some accidental change of ruler or of fortune may bring to the island's shores an entirely different civilization and way of life.[4]

An island can be a bridge or a stepping stone; a frontier outpost or an imperial backwater province; a realm marked by parochialism or a meeting point of civilizations; a place of exile or a laboratory. Islands are thus polysemous spaces. *Like any other kind of space*, their spatial attributes are socially constructed,[5] and are the product of conditions pertaining to time, place and context. As a result, there is nothing fixed or stable in how humans perceive or interact with insular space. Such a reconceptualization of insularity entails a dynamic understanding of changing conditions that are part of a wide spectrum of possibilities ranging from connectivity to isolation, which in turn have no obvious meaning. This is what Lucien Febvre called "the vicissitudes of possibility" almost a century ago, referring to all kinds of spaces ostensibly determined by geography.[6]

Insularités ottomanes, a volume edited by Nicolas Vatin and Gilles Veinstein, represented a pioneering endeavor within Ottoman Studies to see insularity not as isolation, but as a condition of multiple possibilities.[7] The Ottoman world, hosting a large number of islands, provides ample inspiration and historiographical resources to deal with these issues. As a result, several other studies have turned their attention to the meaning of islands as a historical or an analytical category.[8] What is the purpose of revisiting this line of inquiry?

The condensation of time as a result of the bibliographical rate nowadays means that the historiographical and interdisciplinary dialogue on these issues have greatly intensified. The emergence and establishment of spatial history as a new field of study has put space at the center of historical analysis. Beyond programmatic declarations in the form of the "spatial turn," this new trend has challenged strict geographical notions of space. The implications of this research agenda range from the study of global history to questioning how state-centered one's spatial imagination may be.[9] Particularly productive has been the shift away from a terrestrial vantage point that separates land from sea: the challenge to terracentrism has examined aquatic spaces as sites of historical processes, overlooked and marginalized by traditional historiography that privileges land as the stage of history.[10] From a different disciplinary vantage point, archaeology has gone a long way in theorizing insularity, with a particularly rich corpus of works towards understanding island spaces.[11] Closely related are important contributions from classics.[12]

The field of island studies has been consistently studying islands for about two decades now. With its own dedicated journal, book series and various conferences, it has greatly expanded the comparative framework and explored the different conditions that island spaces may entail.[13] Summarizing this corpus of works, there are three main threads connecting the analytical, conceptual and theoretical approaches of scholars within the field of island studies: (1) "nissology" argues that islands need to be studied "on their own terms," rather than from the external vantage point of continental realms and according to which islands are exceptions or paradoxes, among others; (2) islands are microcosms that can provide special insights into larger processes, global or otherwise; and (c) the concept of islandness describes the condition of being an island. Encompassing all these three motifs, the introduction of special issue of the *Geographical Review* stated:

> Islandness is not easily defined. It takes as many different forms as there are islands. Its meaning differs from place to place as well as over time. There are Eastern and Western variants; the islandness of a resident is not the same as that of a visitor; it means something different off island than on. Islandness has taken on a whole new meaning today, since distinctions between islands and continents, once taken for granted, have become muted or dubious. In our time, when people are connected more electronically than territorially, the entire world is becoming archipelagic, with islands appearing everywhere, inland as well as offshore. Cities are "heat islands"; rural areas are "islands of tranquility." Islandness is no longer associated only with water-bound places. The planet itself is now perceived as Earth island.[14]

Firstly, by (rightly) questioning the dominance of mainland perceptions of what insular space is (or is not), island studies sets a quasi-emancipatory agenda from the geographically normative and hegemonic discourse of continental spatial imagination.[15] Yet, such an endeavor falls into the trap of further embedding the very dichotomy that gave rise to this asymmetry, without fundamentally questioning its latent assumptions. The island, *any island*, is not a socially homogenous space. Focusing on the external gaze upon islands as a relation of power is certainly an interesting exercise; yet, it reproduces, if not further embeds, the island/mainland binary.¹⁶ In doing so, it assumes the cohesiveness and uniformity of either kind of space and social relations that exists with it. Lost are the internal differentiations and relations of power within the island itself. In other words, to focus exclusively on external impositions is to forget the internal ones.

Secondly, why should islands constitute a privileged vantage point from which to study any phenomenon or process? If islands are synecdoches and/or metonymies, where does their inherent value lie? Islands *can* be microcosms, but other spatial categories may be too. To think otherwise is to imply an exceptionalism that inverts the one the field of island studies reacts against when "continental" spatial imagination casts insular spaces as parochial, isolated, introvert etc. Regardless of whether one attributes positive or negative value, exceptionalism remains inherent in such conceptualization of islands.

Finally, the semantics of island studies also require addressing. Coining a term in the form of islandness is an attempt to encapsulate the attributes of being an island, literal or metaphoric. Leaving the awkwardness of the term aside, the focus remains on the island itself in an epistemological sense. Rather than a neologism, it would perhaps be more useful to reclaim an existing concept: insularity. The benefit from such an exercise lies in the fact that insularity does not exclusively concern islands, in its literal sense, even if it does so etymologically. More important is the fact that by reconceptualizing this term, one can explicitly add more attributes to the condition of being an island, stripping the exclusivity of isolation that it currently has in its mainstream use. Here is where the reclaiming lies.

The study of islands is not an end in itself. Nor does insularity carry more weight than any other spatial category. After all, why should it offer a privileged access to reality? More than anything else, insularity is an analytical tool, and as such it is a means to acquire a different perspective on the spatial, temporal or conceptual context one may choose to focus on. Just like any other analytical tool, it is a means to move on to different questions, address larger issues and exlore new research agendas. In this sense, what can the study of insularity teach us about the Ottoman world?

This thematic issue addresses three distinct, but overlapping themes that aim to further enrich and contribute to understandings of insularity. Entitled "Conceptualizing Insularity," Part I opens with a compilation of three short essays authored by Spyros I. Asdrachas. He is perhaps the most prominent representative of the *Annales* School in Greece, and has previously published these essays that distill his long experience and engagement with the economic and social history of the Aegean and Ionian islands. A goldmine of information and characterized by synthetic analysis, these essays were originally published in Greek, and are translated and brought together here under the title "Observations on Greek Insularity."

Eleftheria Zei's, with her "The Historiography of Aegean Insularity" presents an overview of Greek and international trends on the Aegean islands and identifies the main paradigm shifts. In doing so, Zei synthesizes an impressive corpus of works, including various published and unpublished doctoral theses mainly in Greek, and situates it with local and international historiographical debates. In her discussion of the historiography of insularity, she traces five main trends which have some temporal and thematic overlaps: (a) the "archipelagic" model of connected versus the "isolated" model of studying islands, influenced by various interdisciplinary trends and persistent throughout the twentieth century; (b) a shift to the study of the local in the 1980s, whereby the study of communities became the focus of attention; (c) during the same decade, a renewed dialogue between historical geography and demography; (d) a concern with the role of islands in the study of the Greece's transition to capitalism; and, (e) the intensification of a dialogue between history, sociology, and anthropology with reference to islands at the turn of the twenty-first century.

Moving on to more specialized case-studies of broader themes and topics, Part II examines aspects of violence and law in terraquous space. Michael Talbot's article "Separating the Waters from the Sea: The Place of Islands in Ottoman Maritime Territoriality during the Eighteenth Century" highlights a little-known aspect of the Ottoman spatial vision for the Mediterranean in their attempt to regulate the activities of pirates and corsairs. In doing so, he illustrates how islands are not unique maritime spaces, and highlights coastal areas as the spatial unit of analysis most relevant for the purposes of his study. Thus, he brings to the fore the notion of littorality, a concept largely overlapping with insularity. The emphasis on coastlines as the meeting point of land and sea shows how islands share many characteristics with the coastlines of continental realms; their only difference is that they are surrounded by the sea. Braudel compels us to return to his work once more: he has a whole section entitled "Seas and Coasts", within which he incorporated his chapter on Mediterranean islands.

Maintaining the temporal focus on early Ottoman history, Cem Mengüç opens up different questions with reference to the issue of maritime violence, in line with the previous two contributions. He draws attention to a rare and important primary source: the historical narrative of Safai describing the Ottoman conquest of Lepanto (Gk.: Nafpaktos; Tk.: İnebahtı) and Modon (Gk.: Methoni) in 1499. Authored in vernacular language and containing first-person narrative experiences of that maritime expedition, it provides unique insights into the day-to-day experience of life on board, before, during, and after the battle. As Mengüç himself argues, this source greatly substantiates the research agenda of a social history of Ottoman maritime space. In this context, islands occupy an important position in Safai's narrative, primarily as sources of water and other provisions. In this sense, we return once more to the question of the overlaps between insular and coastal history, highlighted by Talbot.

In the final section of the volume, contributors turn their attention to the ways the state has imagined and tried to regulate insular space. In her article entitled "Blockading an Island: Collective Punishment, Islanders, and the State in the 'Largest' Island at the End of the Nineteenth Century," Fatma Şimşek relates the story of an incident involving the island of Kastellorizo (Tk.: Meyis). The looting of a half-sunk commercial ship carrying grain is an incident that should raise no eyebrows. Yet, international pressures, as well as a zealous Ottoman enforcement of the law along the lines of a *Rechtsstaat* (law-based state) logic, led to the collective punishment of the islanders through a blockade that marked a radical break with the existing Ottoman spatial imagination of the island, which it practically viewed and treated as connected to the Anatolian coast.

Finally, Kahraman Şakul discusses the Peloponnese during the Age of Revolutions, arguing that for all intents and purposes it should be treated as an island. This may be a valid exercise not only because of the narrow strip of land that connects it to the mainland. In Ottoman bureaucratic parlance, the province was referred to as *Mora ceziresi*, a fine example of the ambiguity between the island/peninsula distinction in Ottoman Turkish. Even if Ottoman bureaucrats may have been saying "peninsula," technically speaking, administrative practice indicated a spatial imagination more akin to that of an island. The same is true of the Peloponnesians themselves, some of whom went as far as envisioning a joined Muslim and non-Muslim confederation as a French protectorate for what they referred to as their *patrie*. Such was the degree of quasi-autonomy from Istanbul, that one observer of the 1770 revolt on the island described it as a "republic of *ayan*, *hakim* and *kocabaşı*, who considered the Ottoman provinces as their family inheritance."[17]

More important than the questions a research project answers are those that it gives rise to. A great many topics were left outside of this volume's purview, while others emerged out of the articles published here. What follows is an attempt to highlight some of these issues as topics of possible future research. For one reason or another, papers in this volume concern islands in the Ottoman Mediterranean. Yet, this was not the only maritory of the Ottoman world.[18] Could there have been a latent assumption that the Mediterranean constitutes the "core" Ottoman maritime space? If so, where else would one seek Ottoman insularities? If we can treat the Peloponnese as an island, to what extent could we do the same for the Crimea? Indeed, such an endeavor would be very interesting given the completely different position the two peninsulas occupied in imperial spatial vision. What about the Red Sea—a recently-rediscovered Ottoman maritime space?[19] To return to the Mediterranean itself, one study has applied the concept of insularity with reference to the Maghreb, an often neglected area of research within "mainstream" Ottoman Studies.[20] Finally, what of Ottoman riverine environments that contain islands? Both the Nile and the Danube are such cases, and it would be worth exploring what sort of insularities existed there.[21] Indicative is the story of the no longer existing Ada Kaleh (Tk.: Ada Kale) island in the Danube: hosting a small ethnically Turkish population in Romania as late as 1972, the island was submerged by the river following the construction of a dam.

Finally, a few words are in order for the individuals and institutions that directly and indirectly made this volume possible. William Blair, the editor of *Princeton Papers*, has enthusiastically embraced the idea of the theme from the beginning, and supported it in every possible way. Anna Papaeti skillfully edited the contributions of this volume, as well as translated one of the three short essays by Spyros I. Asdrachas.

This publication is part of the "Mediterranean Insularities (MedIns)" project (http://medins.ims.forth.gr/), funded by the European Commis-sion's Marie Curie Actions Intra-European fellowship (Reference number: 630030) and hosted at the Institute for Mediterranean Studies, Foundation for Research and Technology—Hellas (IMS/FORTH) in Rethymno, Greece. Elias Kolovos was the Scientific Coordinator of the project, and his energy and enthusiasm were crucial to the successful completion of the project. IMS/FORTH proved to be the ideal host for the project, a true success story in Ottoman Studies in crisis-ridden Greece, in the words of a recent observer.[22] I was particularly lucky to teach a brilliant group of postgraduate students of the M.A. Program in Turkish Studies of the University of Crete; some of the stimulating seminar discussions we had also shaped my ideas. The other members of the Ottoman Studies sector, Marinos Sariyannis and Antonis Anastasopoulos, the then Institute Director Christos Hadziiossif, as well as Apostolos Sarris, Tuna Kalayıcı, Evita Kalogeropoulou, Nikos Tsivikis, and all the IMS staff composed an excellent milieu, of which this volume is also a product.

Some of the articles appearing here were presented at the international conference "Insularities Connected: Bridging Seascapes from the Mediterranean, the Indian Ocean and Bbeyond" held at Rethymno, Crete, on June 10-12, 2016. The conference was co-hosted by IMS/FORTH and York University, Canada, and I was fortunate enough to have Sakis Gekas as a co-organizer. We are both grateful for the financial support of Marie Curie Actions, York University, Levent Yılmaz and AKMED (Koç University Suna & İnan Kıraç Research Center for Mediterranean Civilizations), and the Hellenic Heritage Foundation, Canada. The exchanges between all conference participants informed many of the articles presented here, including this introduction.

That conference was also part of a larger collective networking project. It had built upon previous meetings on the theme of "Islands: New

Theorisations of Insularity in the Mediterranean." Organized by Valerie McGuire, the first meeting was held at the European University Institute in Florence on May 19-20, 2014, and was funded by the Remarque Institute of New York University and the Robert Schuman Centre for Advanced Studies at the European University Institute. The second one was organized by Alexis Rappas at the Maison Méditerranéenne des Sciences de l'Homme in Aix-en-Provence on June 8, 2014. The meeting was sponsored by the "laboratoire d'excellence LabexMed – Les sciences humaines et sociales au cœur de l'interdisciplinarité pour la Méditerranée" referenced as 10-LABX- 0090. It benefited from a state financial assistance managed by the Agence Nationale de la Recherche in the frame of the project "Investissements d'Avenir A*MIDEX" referenced as n°ANR-11- IDEX-0001-02. The stimulating exchanges in these previous meeting, as well as engaging with Valerie and Alexis in the context of that collaboration greatly informed and shaped my own ideas.

Notes

1. Fernand Braudel, *The Mediterranean and the Mediterranean World in the Age of Philip II* vol. 1, trans. Siân Reynolds (London and New York: Fontana, 1972), 148.
2. Edward William Lane, *Arabic-English Lexicon* (London: Willams and Norgate, 1863), 418; Hans Wehr, *A Dictionary of Modern Written Arabic*, ed. J. Milton Cowan, 4th ed. (Ithaca, NY: Spoken Languages, 1979), 147.
3. Katerina Kopaka, "What is an Island? Concepts, Meanings and Polysemies of Insular *Topoi* in Greek Sources," *European Journal of Archaeology* 11, no. 2–3 (2009): 180.
4. Braudel, *The Mediterranean*, 150.
5. Martin W. Lewis, Kären Wigen, *The Myth of Continents: A Critique of Metageography* (Berkeley: University of California Press, 1997).
6. Lucien Febvre (in collaboration with Lionel Bataillon), *A Geographical Introduction to History*, trans. E.G. Mountford and J.H. Paxton (London: Kegan Paul, 1925), 172.
7. Nicolas Vatin and Gilles Veinstein, eds., *Insularités ottomanes* (Paris: Institut français d'études anatoliennes, Maisonneuve et Larose, 2004).
8. In addition to the comprehensive overview provided by Eleftheria Zei in her contribution to this volume, see also Özlem Çaykent and Luca Zavagno,

eds., *The Islands of the Eastern Mediterranean: a History of Cross-cultural Encounters* (London and New York: I.B. Tauris, 2014); Antonis Anastasopoulos, "Centre-Periphery Relations: Crete in the Eighteenth Century," in *The Province Strikes Back: Imperial Dynamics in the Eastern Mediterranean*, eds. Björn Forsén and Giovanni Salmeri (Helsinki: The Finnish Institute at Athens, 2008), 123–136; Evangelia Balta, "The Insular World of the Aegean (15th to 19th Century)," in *The Mediterranean World: The Idea, the Past and the Present*, eds. Kudret Emiroğlu, Oktay Özel, Eyüp Özveren and Süha Ünsal (Istanbul: İletişim, 2006), 97–106.

9. Matthias Middell and Katja Naumann, "Global History and the Spatial Turn: From the Impact of Area Studies to the Study of Critical Junctures of Globalization," *Journal of Global History* 5, no. 1 (2010): 149–170; Barnie Warf and Santa Arias, eds., *The Spatial Turn: Interdisciplinary Perspectives* (London and New York: Routledge, 2009); Kären Wigen, *A Malleable Map: Geographies of Restoration in Central Japan, 1600–1912* (Berkeley: University of California Press, 2010); Karl Schlögel, *In Space we Read Time: On the History of Civilization and Geopolitics*, trans. Gerrit Jackson (Chicago: Chicago University Press, 2016); James C. Scott, *The Art of not Being Governed: An Anarchist History of Upland Southeast Asia* (New Haven and London: Yale University Press, 2009).

10. Peter N. Miller, ed., *The Sea: Thalassography and Historiography* (Ann Arbor: University of Michigan Press, 2013); see the Forum "Oceans of History" in the *American History Review* 111, no. 3 (2006) with the following contributions: Kären Wigen, "Introduction," 717–721; Peregrine Horden and Nicholas Purcell, "The Mediterranean and 'the New Thalassology,'" 722–740; Alison Games, "Atlantic History: Definitions, Challenges, and Opportunities," 741–757; Matt K. Matsuda, "The Pacific," 758–780; Philip E. Steinberg, *The Social Construction of the Ocean* (Cambridge: Cambridge University Press, 2001); Niklas Frykman, Clare Anderson, Lex Heerma van Voss, and Marcus Rediker, "Mutiny and Maritime Radicalism in the Age of Revolution: An Introduction," in *Mutiny and Maritime Radicalism in the Age of Revolution: A Global Survey*, eds. Niklas Frykman, Clare Anderson, Lex Heerma van Voss, and Marcus Rediker (Cambridge: Cambridge University Press, 2013), 5–6; Marcus Rediker, "Hydrarchy and Terracentrism," in *Hydrarchy*, eds. Anna Colin and Mia Jankowicz (Cairo: Contemporary Image Collective, 2012), 11–18; Markus P.M. Vink, "Indian Ocean Studies and the 'new thalassology'," in *Journal of Global History* 2 (2007): 41–62; Rila Mukherjee, "Escape from Terracentrism: Writing a Water History," *Indian Historical Review* 41, no. 1 (2014): 87–101; Jerry H. Bentley, "Sea and Ocean Basins as Frameworks of Historical Analysis," *Geographical Review* 89, no. 2 (April 1999): 218–224; idem, et. al., eds., *Seascapes: Maritime Histories, Littoral Cultures, and Transoceanic Exchanges* (Honolulu: University of Hawai'i Press, 2007).

11. Cyprian Broodbank, *An Island Archaeology of the early Cyclades* (Cambridge: Cambridge University Press, 2000); Kopaka, "What is an Island?" 177–194; *eadem* and Gerald Cadogan "Two Mediterranean Island Life modes, Two Island Archaeologies. Crete and Cyprus: How Near, How Far?" in *Parallel Lives: Ancient Island Societies in Crete and Cyprus. International Symposium Organised by the Universities of Crete and Cyprus and the British School at Athens, Nicosia 30 November–2 December 2006*, eds. Gerald Cadogan, Maria Iacovou, Katerina Kopaka and James Whitley (London: British School at Athens, 2012), 17–33; Burçin Erdoğu, "Visualizing Neolithic Landscape: Archaeological Theory in the Aegean Islands," in *How Did Farming Reach Europe? Anatolian-European Relations from the Second Half of the 7th through the First Half of the 6th Millennium Cal BC,* ed. C. Lichter (Istanbul: Deutsches Archäologisches Institut, 2005), 95–105; A. Bernard Knapp, *Prehistoric and Protohistoric Cyprus: Identity, Insularity, and Connectivity* (Oxford: Oxford University Press, 2008); Helen Dawson, *Mediterranean Voyages: the Archaeology of Island Colonisation and Abandonment* (Walnut Creek, CA: Left Coast Press, 2014); *eadem*, "Deciphering the Elements: Cultural Meanings of Water in an Island Setting," *Accordia Research Papers* 14 (2014–15): 13–26; *eadem*, "Brave New Worlds: Islands, Place-Making, and Connectivity in the Bronze Age Mediterranean," in *Of Odysseys and Oddities. Scales and Modes of Interaction between Prehistoric Aegean Societies and their Neighbours*, ed. Barry P.C. Molloy (Oxford: Oxbow, 2016), 323–342.

12. Christie Constantakopoulou, *The Dance of the Islands: Insularity, Networks, the Athenian Empire and the Aegean World* (Oxford: Oxford University Press 2007).

13. For a comprehensive overview of the literature, see Godfrey Baldacchino, ed., *A World of Islands* (Luqa: Agenda and Institute for Island Studies, University of Prince Edward Island, Canada, 2007); see also the various issues of *Island Studies Journal*; John R. Gillis and David Lowenthal, eds., *Islands,* Special Issue of *Geographical Review* 97, no. 2 (2007); Elizabeth DeLoughrey, ed., *The Literature of Postcolonial Islands*, Special Issue of *New Literatures Review*, 47–48 (2011).

14. John R. Gillis and David Lowenthal, "Introduction," in *Geographical Review* 97: 2 (2010): iii–vi.

15. See Godfrey Baldacchino, "The Lure of the Island: A Spatial Analysis of Power Relations," *Journal of Marine and Island Cultures* 1 (2012): 55–62.

16. For a critique the islander/continental binary see Lisa Fletcher, "'…Some Distance to Go': A Critical Survey of Island Studies," *New Literatures Review* 47–48 (2011): 20–23.

17. See endnotes 1, 3 and 4 of the relevant article.

18. I borrow the term from the excellent presentation by Benoît Bérard, "From

Landscapes to Seascapes, from Island to Archipelago: How to Redefine the Pre-Columbian Territories and Maritories in the West Indies," unpublished paper presented at the *Insularities Connected* conference (see above).

19. Alexis Wick, *The Red Sea: In Search of Lost Space* (Oakland: University of California Press, 2016).
20. Brent D. Shaw, "A Peculiar Island: Maghrib and Mediterranean," *Mediterranean Historical Review* 18, no. 2 (2003): 93–125.
21. I am indebted to Dimitris Giagtzoglou for this idea. "Observing the activities of the officials who were part of the Ottoman administrative and military structure, [...] one would say that this is not a river but an Ottoman province, while references in the documents on islands and fortresses seem to transform the river to a sea." Dimitris Giagtzoglou, "Όψεις της στρατιωτικής και οικονομικής ζωής στον Οθωμανικό Δούναβη: τέλος 16^{ου}-αρχές 17^{ου} αι.," [Aspects of Military and Economic Life in Ottoman Danube: End of Sixteenth-Early Seventeenth Century], Paper submitted for TOY 186: Ottoman Environmental History as part of the Postgraduate program in Turkish Studies at the University of Crete.
22. Michael Ursinus, "Ottoman Studies Triumphant: The Success Story of Rethymno, Crete," *Byzantine and Modern Greek Studies* 40, no. 1 (2016): 89–98.

PART I

Conceptualizing Insularity

Observations on Greek Insularity

SPYROS I. ASDRACHAS

I
The Greek Archipelago: A Dispersed City[1]

"The city—locked within its walls—does not exist," concluded Ruggiero Romano at the close of the discussion on the city in Mediter-ranean history. He continued by stating that

> Historians are well aware that the city is a place which makes its own space; that is, its own internal values give it the tendency to construct around itself a living space which may be extensive to a greater or lesser degree. One might think that this is the old and much-discussed urban-rural relationship. But no. We are dealing here with something much wider: with the fact that modern historiography (really modern historiography, that is) has entirely abandoned the concept of the "isolated city" (*isolierte Stadt*) of von Thünen, although of course the idea continues to be useful when talking of the economy. The past (and the present) of a city ought not to be seen as covering only its urban space, in the strictest sense of the word, and its immediate surroundings: the walls of the city cast a longer shadow than might at first be evident.[2]

So much for the outward-facing, centralized city. But there are dispersed cities, too—cities which are aggregates or networks, and among them is the island complex of the Aegean. The text which follows attempts to examine these questions and become part of the appropriate historiographical viewpoint.

The Greek Archipelago, that unifying sea or plain-like expanse, has always been a focal point for the reception and transmission of cultures, throughout pre-history and recorded history, and has apportioned itself between isolation and a system of constant connections. Without a center of its own, it has forged multiple links between the lands to its East and West; its fragmentary nature allowed its individualized organizational forms to survive throughout the long years of Ottoman rule. Those forms—each a retreat into the self of the Archipelago—were typical of urban structuring and were for the most part formulated during the period of western rule. The community was founded on fiscal solidarity and organized in such a way as to tend to stabilize social mutuality, which in turn permitted the perpetuation of internal balances imposed by two overriding factors: the disproportion between human and natural local resources, and a consequent dependency from the point of view of food supplies.

Not all coastal zones and not all nuclei of habitation are by nature the links in a perpetual chain of structure; the sea may provide opportunities for the continuity and unity of human space, but its cliffs are a frontier calling towards isolation which exists, but does not dominate.

Although the Archipelago is crisscrossed by mercantile and military fleets, it does not make its every island a waystation. Sometimes the crossing is made almost without a halt, as in the case of the journey of the Baron de Saint Blacard, who, in the winter of 1537, sailed from Euboea to Istanbul calling only at Chios. The harbors on the island are laid out according to the needs imposed by the safety of their little fleets; small bays protected by steep and fortified ground formations within which the town, too, nestles, it also being most frequently constructed as a defensive nucleus. Nonetheless, not every fortified point commands the sea, and the agglomerations of habitation with which the fortified points go to make up a whole do not constitute an equal number of points at which maritime movement could be received. They are part of the hinterland, as is the case in the Cyclades with the castle complexes of Kea, of Kimolos, and of Emborio on Santorini. Along with settlements of this kind are others lying on the coast but built on the top of cliffs inaccessible from the sea, with which they communicate through other, more distant, anchorages. Two more types of fortified settlement can be found, in direct relation to the sea: this direct relation springs from their sites, which are either on heights crowning accessible inlets or bays, or on low outcrops of ground in gulfs or inlets or along the shores of straits. These last two types of settlement can be found in the Cyclades on a number of islands, Syros, Sifnos, Ios, Serifos, Andros, Naxos, Mykonos, Paros, and Antiparos.

Thus from the point of view of inhabited areas the way to the sea is both open and closed: this does not mean that these forms of inhabited area reflect an equal number of economic micro-systems, whether self-sufficient or transactional—the latter in the context of the connections between the islands or between the islands and the mainland. This is because the Greek Archipelago consists mainly of small islands, and depends on external input to fill in the gap left by inadequate agricultural production. With the exception of some of the larger islands—Crete and Cyprus foremost among them—most of the others suffered from the disequilibrium between their sources of wealth and their demographic strength. This is even more evident in the case of an island like Crete, whose economy was above all agricultural. Yet even Crete, as early as the Venetian period, was dependent on external sources to meet its needs in wheat—Methoni and Koroni supplied Chania; Thessaloniki and Euboea supplied Candia (present-day Herakleio); and outside the area of Venetian conquest, the coast of Western Anatolia and Alexandria provided extra supplies for the islands of the Duchy of the Aegean.

The response of the islands—and especially of those living on the small islands—to the disproportion between natural sources of wealth and demographic strength consisted of occupational and productive specialization: fishing, shipping, and trade constituted a privileged area of activity which allowed this disproportion to be overcome, while the local economies were being monetized and a system of communications was established. One should not believe that shipping was exclusively a response to a demand for sea transport facilities arising at the points where there was surplus production and at which products were directed into long-distance transactions: it was more of a response to the demands of local economies and their manner of specialization.

The fishing boats give proof of this in the islands which became specialized in sponge-fishing: Kalymnos, Symi, Chalki (Dodecanese), and above all, Kastellorizo. There can be no doubt that voyages which were not related to specialization and local shortages of food had created flourishing commercial fleets—the examples of Hydra, Psara, and Spetses during the continental blockade have much to tell us, and the same is true of the fleets of other ports: Messolongi, for instance, during the second half of the eighteenth century, and the little Aegean island fleets which secured constant communication with Venice. From the vantage point of the economic continuity in the Aegean, however, of prime importance are the connections of a different kind: those between islands.

This economic unity transforming this liquid world into an urban complex was not simply due to the fact that the islands succeeded in maintaining long-term maritime contact between the two halves of the Mediterranean, which intensified from time to time due to changing circumstances. Nor was it just that all routes from the Eastern to the Western Mediterranean had to pass through the islands. If the complex existed, then the main reason for its existence has to be sought elsewhere: in the constant to-ing and fro-ing of people, goods, and ships from one island to another, in an economic osmosis, at least insofar as this is manifested through multiple connections which are simultaneously internally and externally-directed.

But before going on to present some examples related to these internal communications and interdependence, I would like to provide a reminder of a number of fairly obvious facts. The Greek Archipelago was always a part of wider state formations, in this case, of the Ottoman Empire; in addition, the different rulers it experienced brought about its administrative fragmentation. Yet, in all cases the islanders felt that they were part of a single administrative and state entity. The fact, however, that the islands were under the control of different rulers, and that they had contacts with both halves of the Mediterranean—with all the cultural offshoots which this contact implied—meant that the feeling of membership in an entity went hand-in-hand with the perpetuation of individualities. This situation was not without its interruptions, when political circumstances dislocated the interdependence, as happened with the Cretan war which broke off links between Patmos and Crete and Venice, and emptied

Patmos of its non-monetary goods—"silver, gold, and pearls, they took them down to the galleys in sacks." Sometimes, too, the islands' position within a state came into contradiction with the obvious spatial facts: for the inhabitants of the Dodecanese, the "journey inside" was the trip into the Adriatic, while that along the shores of Western Anatolia was known as the "journey outside."

The Ottoman Empire had grouped most of the islands (the Cyclades, the Sporades, the Saronic Gulf islands, and those of the Gulf of Argos, together with three of the Dodecanese) into one administrative unit. Nonetheless, the system of communal administration imposed Istanbul as the main point of political reference: it was there that the representatives of the island communities sought to resolve their collective affairs, each on behalf of his home island but also in conjunction with measures taken in respect of the other islands. More partial administrative units which were at the same time tax regions, reinforced relations between islands forced to bear the same burdens. Thus Naxos and Paros formed one province of the old Duchy of the Aegean, under the same bey, as did Andros with Syros and Milos with Santorini. "Andros and Syra are one" —and they were for the Ottoman fief-holder of the two islands in the eighteenth century. Of course, this did not mean the formation of a city scattered across the waters, but it did strengthen mutuality. This mutuality can be seen more or less everywhere: in the islands' obligation to maintain vigilance over the seas and to keep in touch with fires by night and pillars of smoke by day; in the 1812 Kythnos communal regulations, which forbade certain forms of behavior as likely to shame the islanders in the eyes of others: "lest we become the laughing-stock of surrounding islands;" in the jokes which one island told about the other; and, chiefly, in the distribution of roles, of which we shall have more to say later on.

If each island looked to its neighbor night and day, it was because they shared the same fate and trembled at the same fears. Among those fears, the most notorious were the pirates and corsairs, however beneficent they may have been for some of the islands. There were other fears, too, of a daily nature, which had nothing to do with piracy or war but which nevertheless left their mark on the community of spirit, on the most intimate beliefs scarring the landscape of the soul: the vampires, wondering freely almost everywhere impeding communication between the two worlds as it is manifested in the worship of the memory of the dead. This is a game of submission and complicity played between the human and the divine and confirmed by a multitude of churches and monasteries and represented by a numerous clergy and a well-organized monastic mechanism in the large foundations such as that of St. John the Theologian on Patmos. Here, spiritual hegemonies testify to the distinctive features of a society that, albeit roughly similar, is in places the product of multiple ethnic and religious blends and constantly reinforced by internal and external migratory currents. These hegemonies varied in intensity, and were even differentiated from each other from place to place. The Christian Orthodox composed the essential part of the population of the islands, apart from Naxos, Tinos and Syros where Catholics were the majority. Elsewhere the latter formed aristocracies which, despite mixed marriages, were able to invoke their western origins and confessional denomination to set themselves apart from the rest of the population. Indeed, on Naxos at one point they became entrapped within a decaying atavism corresponding to the island's arthritic economy, which had turned in upon itself and relied on income from land rent, although it did not cease for this reason to provoke constant social tension.

Although the society of the islands was, broadly speaking, similar, social forms varied. If the agricultural population was omnipresent, its demographic significance differed from island to island. The same is true of the city/village dichotomy, which was pronounced in Rhodes with a clear dividing line that the walled cities impose: their gates were closed each evening against the coming and going of the villagers. We find a similar phenomenon in all the protected areas, but here the separation was due also to the economic features of the cities. Proof of this exists in one of the chronicles of the late sixteenth century, which describes the typical famine/epidemic cycle: the classic mechanism analyzed by Wilhelm Abel. As everywhere, the starving villagers attacked the walls of Candia; the city closed its gates to them and the nobility took refuge in its fortified and well-supplied villages, leaving the *popolanoi* (plebeians) to be decimated by the plague. As a result, manufacturing suffered from a shortage of working hands, and wages rose. These were urban phenomena that are not as clearly discernible in small islands: on barren Serifos a population decline in the late seventeenth century did not mean a consequent restructuring of land ownership and a rise in family income. There are other similarities, too, relating to urban stratification: a class of nobility (*archontes*) in Catholic-held islands, and the remnants thereof during Ottoman rule; where merchant shipping created fortunes, the population was classified with precision on the tax registers; householders (*noikokyraioi*), priests, and laborers (*kopiastes*). Social stratification also revealed itself in more paltry symbolism: on poverty-stricken Skyros, which had no shipping to speak of, only the "noble" families had the privilege of owning certain types of copper utensils, which were marks of wealth amidst the general impoverishment. But these social and economic similarities, the component parts of which are unequally distributed, are not sufficient by themselves to make a dispersed entity into an urban complex. They do, however, show that each human community on the islands sought to maintain its internal structure, the equilibrium, one may say, of its own totality. This was an essential pre-condition for individual totalities to become part of a general whole that was the dispersed city.

Fulfilling tax obligations was the primary condition for the maintenance of the equilibrium in any society. Since tax liability was collective, in order to sustain its own existence each insular society had to operate a system of solidarity that would assess and determine individual taxation according to wealth. This practice reinforced social stratification and led to the sensitization as "citizens" of those who belonged to the upper strata: merchants and ship-owners; nobility of Latinate origin; and even local notables, including the clergy who distinguished themselves because they owned more than the moderate agricultural wealth of the islands. For the collective body, as expressed through the codification of custom, the ideal equilibrium consisted of respect for the "ancient" rules and regulations, which stabilized the perpetual ownership of goods within the family circle. In the succession of heirs, descent through the male line was usually the rule. Yet, where the right of primogeniture was enforced, it did happen that women were favored, as was the case in Karpathos. This osmosis at the level of custom, despite the sporadic occurrence of the right of primogeniture, constituted one of the elements in the cultural unity of the islands; seen from the vantage point of the mechanisms regulating agricultural production, this unity of custom was a reflection of the disequilibrium between the islands' demographic strength and the natural (agricultural) sources of wealth. Indeed, the lack of arable land led to an extreme division into small-holdings, which was countered by the restructuring of family fortunes either by the right of primogeniture or through the exclusion of women from distribution of the paternal fortune (*materna maternis, paterna paternis*). These distinguishing features of the islands merge into an integrated social and economic community; this integrated community was being confirmed within every island by personal institutions, communal authorities, and codified local customs. In order for all the above to exist, the islands had to be open and to communicate.

As mentioned above, the islands of the Greek Archipelago counterbalanced their shortages in agricultural produce by alternative forms of production, namely the provision of services. They were, then, open and communicating societies. Their populations were formed in relation to migratory movements, such as, for instance, Latinate and Albanian settlement. At the same time, there was also a reverse trend from the islands to the Western Anatolian coast: the settling of Naxiot wine-growers and barrel-makers in Izmir in the mid-eighteenth century; the establishment of people from Syros, Kythnos, Serifos, and Amorgos in Istanbul and on the Western Anatolian coast; or moving to the latter regions from Chios, Lesvos, Samos, and the Dodecanese. Migration is also apparent from one island to another. Names, among other evidence, can be used to prove this: of the 38 indicating specific origin found on Kythnos, 21 are from Aegean islands; in 1673, fifteen percent of the population of Patmos bore names of local origin. Whenever we have available dated records of a representative part of an island population, we can see the internal Aegean "diaspora." Here, the contribution of Crete stands out. The same is true with toponymy. Inter-marriage was one of the causes of these movement: "we have many women of Tinos here, too," wrote the representatives of the local community (*koinon*) of Syros in 1781. This was, in part, due to the bonds between Catholic communities. But let us return to maritime communications and the economic links which they brought about.

There were two types of communication: the first was by means of shipping and did not involve the inhabitants of the islands. Fleets or individual ships would stop at the harbors or inlets of the Archipelago to take on supplies or to shelter from bad weather. In this instance, spatial and especially economic continuity involved only those who travelled and not those who lived on the islands at which they called. The other form of communication involved participation on the part of the islanders themselves as either the direct agents of shipping or as its clients. The communication generated by these means could be internal, external, or both. The economically continuous space is a correlation of this type of communication; it is also in correlation with this economically continuous space *for the islanders themselves* that we much define as the dispersed liquid city. The direct links between a port such as Chios and external centers by means of the arrival of individual ships or convoys did not sensitize the other islands, and the port of arrival, despite the transactions taking place there, did not become a pole of attraction in the form of caravanserai or a temporary city as in the case of fairs. This does not mean, however, that the presence of ships in a harbor securing extensive movement between the two halves of the Mediterranean did not provoke the interest of the island merchants. It simply means that the presence of ships did not set off a reaction equivalent to that of the cities of caravanserais or fairs. In other words, these were important seasonal events localized in time and with sufficient power of attraction to orientate the movement of all the islands' commodities towards the one pole.

The urban complex of the Archipelago was not only a corollary of transactions eventually resulting in a market outside its own area; this market constituted a factor to the extent to which local production was subject to long-distance transport transcending the mainland. It should not be forgotten that the islands did not indulge in long-range shipping: in about 1813, Kastellorizo may have sent 30 sponge-fishing boats and some 450 seamen in search of sponges off the coast of the Maghreb, but Serifos had to wait for ships from Hydra to load its grape must at predetermined prices. This last example is indicative of the way in which economic unity was achieved: the communal authorities of the two islands defined the terms on which the one community absorbed the products of the other. Grape must from Serifos was to supply the market of Hydra, and each sale was to be made on the terms of the agreement drawn up by the communal authorities of the two islands. This agreement stipulated that sale elsewhere could only take place after the requirements of the purchasing community had been met. Restrictions of this kind were innate in the functioning of the communities, and can be seen at various levels, perpetuating the right of "preference." For example, this is visible with regard to the disposal of land, provisioning, or to the way in which the market operated, itself aiming at serving immediate consumer needs which limited the range of the merchants.

All the evidence makes it plain that the economic unity of the islands was dictated by their interdependence in relation to foodstuffs, and did not concern external trade. Nor, with the exception of wheat, did the composition of the inter-island market reflect that of the external market. Let us examine some examples of this.

By the mid-seventeenth century, Kimolos had lost the wealth which the pirates had brought. Its crops were divided between barley and cotton, and its consumer needs were met by importing wine from Milos. Its only exportable manufacturing product, cotton socks, "supplied the neighboring islands." Milos, in turn, was during the time of the pirates "the great bazaar of the Archipelago." Its booty, at bargain prices, made the fortune of the "middle classes," which "sold it off at a profit." When the good years were past, the Meliots were content to export their hand mills to Istanbul, Egypt, the Peloponnese, Zakynthos, Cephalonia, and Ancona. In Sifnos, too, they drank the wine of Milos. A fertile island itself, Sifnos produced all the cereals it needed and made cotton cloth. Thus it was able to absorb part of the cotton production of the surrounding islands; its straw hats were to be found throughout the Archipelago. It also exported honey, wax, onions, and sesame seeds. There were also other forms of movement beside commercial transactions: pilgrims, for instance, who came "to pray." In Andros, the locals worked in silk and built ships. The islanders controlled the means of transport and themselves exported their products to Chios; in the villages they made baskets which were exported to all parts of the Archipelago. So there was some manufacturing in the villages, too: most of the villages of Chios made textiles for local consumption and for export.

Santorini produced good wine which was exported to Chios, Izmir, Istanbul, and Crete, but lacked cereals. It thus sent its boats to buy supplies from Amorgos. Santorini also imported wood from Folegandros, which consumed its wine. Patmos had commercial communication with the coast of Western Anatolia and the Gulf of Venice, but also with Egypt,

Crete, and the Aegean islands, of which Samos supplied it with timber from building, Santorini with sails, and Chalki and Kalymnos with sponges. A detailed sample drawn from the collection of the *koumerkio* (Lt. *commercium*; Ott. *gümrük*) duty in Patmos in 1783 illustrates the inflow from one island to another: of the islanders which carried goods, Samiots brought iron, cotton, onions, cheese, wine, and wax; Kasiots brought soap, hides, sponges, cotton, canvas, and salt; Mytilenians brought olive oil; Hydriots brought linen and, of course, wheat; Chiots brought pots; Symians brought soap, hides, sponges, cotton, canvas and salt; Cretans brought oil and soap; Nisyrians brought pots; Parians brought wine; Ikariots brought charcoal; Kalymnians brought sponges; Mykoniots brought oil; Cephalonians brought flax; Kastellorizians brought rafters and planks. The list could go on: in 1747 we find sailors and merchants from Santorini, Amorgos, Astypalaia, Skopelos, Leros and Tilos—a microcosm weaving a net of communications from one end to the other in the acentric liquid city of the Aegean.

The production of textiles, sails, socks, and gloves constituted an area of manufacturing that is conducive to internal trade. In the second half of the seventeenth century, Chios imported 9,360 kg of silk from Tinos and the Thessaloniki region, as well as 8,000 *scudi* worth of silk from Samos. The cloth made in Chios was not destined for the islands, where raw materials were sought after, but for Istanbul and Izmir. The flow was not always reciprocal, and not at all levels. There were inequalities and diversifications: the merchants of Chios became the "conquerors" of the Levant in the nineteenth century as traders and financiers. What is important, however, is that all the islands, with their inequalities of demography and economy, were forced to develop this form of communication amongst themselves. If there was a ranking of magnitude, it was not due to one island dominating the other by compelling it to produce only agricultural commodities; it was a result of the fact that in an agricultural production *par excellence*, merchants were able to primarily integrate a part of the agricultural surplus into large-scale commercial networks, as well as, secondarily, a section of manufacturing production. Some products were able to take part in this game: a case in point is that of the silk imported to Chios to cover the shortfall in its own production. But the real dependence of the islands is to be sought elsewhere: in their shortage of foodstuffs. From this point of view, imports of oil from Mytilene and Crete, of wine from Psara and Mykonos, and of wheat from Western Anatolia provide a better illustration of the dependence of Chios on the islands as a whole, and the dependence of these islands on the cash flowing in from Chios.

Interdependence led to a general complementariness of the islands' economy; even a complementariness of economic roles. It should not, however, escape us that both of these assume links with the mainland, which both provides and receives. The archipelago is acentric, hence not free of localized tensions: Rhodes, for instance, with its multiple internal and external connections, which were, however, loose throughout the eighteenth century; Crete, with its lasting connection to Marseilles; even Kos, which sent the products of Karaman to Alexandria; or Hydra and its leading role at the time of the continental blockade. The list could be a long one.

As we have already noted, some islands specialized in one main activity—the sponge-fishers of Kalymnos, Symi, Chalki, and Kastellorizo are typical instances. There were also other specializations or divisions of labor: the ship-owners of Mykonos, acting on their own behalf, transported timber for construction from Mount Athos to Alexandria, and on their way back carried coffee and rice, not only for the islands but for the mainland (the Peloponnese) as well. Mykoniots, known to be the best sailors in the Archipelago, found work in the fleets. Meliots were considered the best pilots; Symians the best divers, much sought-after for the recovery of sunken ships; and Parians were reputed to be the best arbitrators of disputes.

One could find endless examples of the direct and multiple links between the islands and the distribution of roles. One could insist equally well on the inequalities and diversifications which occurred in maritime space and which turned Psara, at the start of the nineteenth century, from an ordinary island into an enterprise managing the economies of the other islands. One could also qualify the somewhat exaggerated positive image which might arise from a description of a mechanism as a whole rather than the systematic recording of the ways in which it was put into operation. Let us devote a little space to this positive image.

We are discussing here a maritime world with communications which were not always easy or uninterrupted, for the Aegean was not an easy sea to navigate throughout the year; nor was it always a peaceful region; nor were all techniques equally able to control a ship at sea. On another level, commercialization too fell foul of that internal isolation—which was not, nevertheless, dominant, as mentioned above. A classic example is that of eighteenth-century Chios: "most of the mountain-dwellers will trade only by barter, and since they are forced to consume their own wine, because they cannot transport it, they would be penniless if they had not had the foresight to breed flocks." This testimony is illustrative of two issues: on the one hand, it shows that trade was impeded by lack of communications in the island's hinterland; on the other, it shows that minimum monetization can even penetrate human environments where barter is dominant over monetary exchange.

To conclude, it will never be possible to quantitatively and tangibly demonstrate the significance of internal trade relations and the consequent hierarchies within a while that is also rooted in an agricultural economy, which, on the larger islands—but not exclusively—constituted the heart of the whole economy. Our problem, however, lies elsewhere: whether "a liquid space from an expanse of sea, from which extrudes a host of islands (with their towns, villages, and settlements, their ports, internal ship-lanes akin to roads, and faint differences), may as a whole take on the form of one single city dispersed in space." Of all these elements, the one which creates cohesion is the network of internal communications: through them the whole reaches equilibrium. This whole is a broad conglomeration with communication and interdependence, where one-way connections are not dominant. And is that not a description of a city?

II
The Islands[3]

A part of the Greek lands is directly dependent on the sea; the rest is indirectly so, through multiple mediations. Together they compose a multifaceted nexus of economic and social formations that end up, again through a system of dependencies of a different scale, in a partially geographic and partially economic major dependence on the sea. These dependencies are not unilaterally defined by geographical coordinates. During the early modern period, the whole of the Greek lands lived under foreign rule, Ottoman and Western, the latter mainly Venetian. It was, therefore, part of a network of naval powers, which nonetheless brought with them institutions, composed social formations, and depended on economies grounded in the most predominant, timeless constant of Greek history: the agricultural economy, whatever this generalized term denotes. This economy was omnipresent, defining the limits of monetization, partly driven through internal and external sea communications. This space consisted of zones of equally numbered internal networks, characterized by their own communications. At the same time, it was perforce affecting long-distance movements: thus, the most dynamic part of the space is people.

Islanders or continentals, people were also subjected to movements, forming part of different social groupings. Next to the overwhelming majority of land cultivators, whose traits were by no means unified, we can discern the clear characteristics of several other categories: city dwellers, such as men of the market and manufacturing, men of trade and finance, as well as cultivators growing their vineyards around the city or engaging in the monotonous cultivation of cereals. There were also people of the sea, sailors and merchants accompanying their goods, navigators, fishermen and sponge divers, shipyard workers, naval fleet crews, or men of land or maritime warfare, who occasionally switched roles.

Western rulers lost power due to Ottoman expansion originally in the islands, and later in certain coastal areas. First the Aegean islands, and then Cyprus and Crete were incorporated into the Ottoman Empire. Alongside architectural works and fortifications, other traces betray the large-scale cultural osmosis that occurred through the Latin conquest. Other elements, however, would disappear, such as Cypriot poetry and Cretan theater that equally sprang from the Renaissance. Any Western continuity that persisted in the Ionian islands meant that societies there would be composed through a borrowing of elements, adjustments, and eventually phenomena of cultural syncretism. As a result, intensive phenomena related to education would not assume a form operating within a national ambit before the end of the eighteenth and the nineteenth century. Whether intensive or expansive, cultural diffusions and their ensuing assimilations were the result of a broader diffusion, taking place in the eighteenth century initially in the Mediterranean and later beyond it. This was preconditioned by the mobility of people, goods, and ideas, rooted in the principle of Mediterranean universality. Even though the Mediterranean was divided in two worlds, it was at the same time characterized by continuities, by a shared technical language, and by interspersed common references. This was the case even in its eastern part, which gradually fell under Ottoman rule, becoming the edge where the pulsing of its entire financial mechanism eased off: where trade transport and everything that came with it on a social level, created a differentiation in scale with regard to economic, social, and intellectual values, even if it did not entirely change the economic mechanism. In areas under Ottoman rule, we can observe these phenomena with relative clarity from the second half of the eighteenth century.

Throughout the sixteenth century, the Eastern and Central Mediterranean seas were disputed seas. Bearing no immediate results, the Battle of Lepanto would mark the limits of Ottoman supremacy, which would nevertheless eventually unify the region in the late seventeenth century. The short-lived Venetian reconquest of the Peloponnese would also come to an end in the early eighteenth century. However, the Eastern Mediterranean will not cease to be an arena of military conflict throughout this, and the subsequent period. Up until 1821, Greek island populations lived under conditions of war, but were not defined by it. They experienced war as victims, as crews of naval fleets that clash in battles. Sometimes they even benefited from it; indeed this is the case after 1770, and especially during the continental blockade at the time of the Napoleonic Wars. They also experienced this game of political rule through another, endemic kind of war: the *corso*. The boundaries between the *corso* and piracy are blurred, but outcomes are the same. They were beneficial for some islands in the Aegean, essentially as pirate refuges as well as markets for the spoils on offer. The main agents behind the *corso* were Westerners, the Knights of the Order of Saint John being the most prominent. Towards the opposite direction, there were the Barbary States. However, Greek island and coastal populations were also involved in the violent acquisition of goods using the same methods. Indeed, they were recruited as pirates by Christian powers up until the beginning of the nineteenth century. Nevertheless, the populations' relationship with the sea for the most part was not defined by war and its effects, nor by fixed geographical terms.

Not every island and coast points to a maritime opening. Agricultural constants call for isolation, just like threats coming from the sea lead to settlements defensively situated away from the coast. However, there is no absolute typology regarding this. If in one case settlements spread in parts of coastal areas that cannot be accessed by the sea, in other places they are linked to the sea through an outpost settlement. Or they choose more accessible and protected areas, ship way-stations or places that are also destinations of trade transport. The relationship with the sea, which is both socially and demographically selective, is defined by more fixed historical coordinates that belong to the *longue durée*.

Let us start with one of the most extreme of these coordinates: the disproportion between demographic potential and local natural resources. On the one hand, it directed the population to maritime professions, shipping, and fishing; on the other, it was conducive to simple and temporary immigration, as in the case of the Aegean islands. Alongside their naval activities, they sent their population surplus to Anatolia or to the melting pot of Istanbul. Deficiencies in natural resources, particularly so in food supplies, dictated communications between the islands as well as a distribution of roles within the island complex. An island channeled its goods to other islands in exchange for what it needed. This movement was twofold. It took place among the islands, as well as between the islands and the Greek and Anatolian mainland. If more or less all islanders went out to the sea, not all of them had the same role. Some islands had a supremacy in shipping. Indicative examples are Hydra during the continental blockade; the port of Messolongi at some point during the eighteenth century; or Candia's direct connections with the Gulf of Venice during the sixteenth century. Nonetheless, beyond this alternate interrelation of maritime forces, there were additional distribution of roles: some are more permanent, such as excelling in sponge fishing, underwater ship-repairs, and the recovery of sunken boats. More temporary ones include, for instance, the ships of Mykonos going back and forth from Mount Athos to Alexandria, and from there to the islands and the Peloponnese.

On their way to Alexandria they were loaded with timber, and on their way back with coffee. Alongside these, there were other specializations. Sometimes these were defined by local manufacturing and cottage industry: socks from Patmos; cloths made of thick wool from Santorini; hand mills from Milos; silk fabrics from Chios; straw hats from Sifnos, and so forth. Other times they perpetuated nautical knowledge: the pilots of Milos, the famous sailors of Mykonos, and the divers of Symi. All these specializations are discharged to the sea; they were defined by the sea and its routes.

As previously mentioned, these routes are both internal and external. They connect islands between them as well as with ports of the mainland, and both with the two parts of the Mediterranean. Though sailors are part of these inter-Mediterranean connections, they do not determine them. At times, as for instance in the eighteenth and the beginning of the nineteenth century, they assume a prominent role due to political conjuncture. The rule, however, is for them to have precedence in the passage to the Eastern Mediterranean, and later, in the nineteenth century, in transports via the Black Sea. Western ships dominate commercial navigation, western ships were predominant before the sixteenth century, and from then on northern ones occupied that role. This was an economic supremacy of the cities of the western Mediterranean at first, with northern ones taking the lead from the end of the sixteenth and through the seventeenth cen-tury. Western cities, most notably Marseilles, regained a position of prominence in the eighteenth century following the British retreat. In this game, locals assumed multiple roles, all of which converged on subjugating to commercial capital the main source of commodities: agricultural production and the goods of direct production. This is because exports to the West were primarily based on agricultural goods: cotton, wool, cereals, and oil. Other goods, in smaller quantities, included manufacturing products, yarn and hand woven fabrics, and rawhide leathers. By contrast, Western imports in the Eastern Mediterranean essentially consisted of treated products, products of Western manufacturing. Coins were also imported, another kind of commodity among others, which was also used as a general equivalent benefitting from the debasement of the Ottoman currency, that is its reduction in silver consistency. The circulation of money was slow, and was dealt with by this influx of foreign coins despite the fact that selling western products brought lucrative profits.

In this kind of exchanges, which can be clearly observed in the eighteenth century, locals assumed many roles. Some were based abroad, in Italian as well as Central European cities, and formed trade companies with their compatriots from the Greek regions of the Ottoman Empire. Others were permanent Ottoman residents, traveling merchants in East and West, forming mostly short-lived companies accompanying their goods. Others stay put, entrusting the transport of their goods to sea carriers. Occasionally the distribution of commodities is taken over by these carriers, captains, ship-owners or both, and not by a partner abroad or a representative. Other times locals had a mediatory role. They became the intermediary between producers and foreign merchants based in ports of the Levant which had no direct access to production. To the extent that locals were in charge of trade within the two parts of the Mediterranean, they acted just like foreigners. That is, on their way back they brought western products as well as money, sometimes mainly money, for instance from Central European markets. This is an old system, which is intensified from the second half of the eighteenth century onwards: either through sea or land, it is part of a wider transport, whose significant dates are trade fairs at the Balkans, and even big fairs in Leipzig and Italian cities, such as Senigallia and before that Lanciano and Recanati. Sea transport plays a big part in this. Also important is the presence of Greek ships; their significance is certainly not measured by moments of favorable political conjuncture.

In Venetian territories, transport was defined by the policies of the ruling power which directed all routes towards the metropolis and determined their largely agricultural production. From the sixteenth century onwards, certain reductions in custom duties or some relevant minor privileges in Crete and the Ionian islands led to a loosening of restrictions of commercial freedom. A corresponding phenomenon in the hinterland of the islands was concern for agriculture, even if this was still directed towards monoculture. A merchant navy notwithstanding, particularly in Crete, there was no remarkable shipping record in Venetian territories; the same applied to trade. However, Venetian rule did not allow everyone to benefit from its agricultural policy, particularly cultivators. Even if the exclusivist trade policy of Venice did not favor the (external) trade of its possessions, it nevertheless instigated a broad range of social and cultural differentiation: the emergence of a local urban lifestyle, with intense social stratification; cultural contact is reflected in cultural formations in art and poetry, promoting a harmonious westernization that brings out intrinsic, "national" elements; Venice and Padua received a lot of islanders who were incorporated in the predominant culture, while maintained their own cultural differences. When cultural difference was marked by inflexibility, as in the case of religious art, cultural contact eventually led to the adoption of western tropes to the degree to which the latter did not undermine the context of reception. It would take time for a transition to a new kind of art in the Ionian islands; Panagiotis Doxaras would be its most prominent figure. By contrast, in poetry, alongside assimilation through the adoption of the Italian language, cultural contact and cultural formations, completed in Crete, were dominated by local cultural realities that integrated corresponding elements; these may have been borrowed, but would later develop into something entirely new.

Let us return to the constants that defined the relationship with the sea. The vastness of the Ottoman capital instigated internal and external maritime movements. It also caused corresponding land movements, even migration. Temporary or permanent, obligatory or voluntary, migration reinforced the demographic weight of Istanbul, as well as professional diversification. Of these movements, those marked by an economic character addressed the consumption needs of a large city; the most distinguished of these movements served its need for cereals.

Thus, in the context of a state-directed provisioning policy for the capital, direct communications were developed between the capital and cereal-producing areas. Such communications were made possible by sea and river transports: the Danubian Principalities, the Aegean coasts, Thrace, the Black Sea, and the province of Bursa. The same system supplied cereals to other areas in need. However, the existence of communications and the circulation of goods did not predetermine the mobilization of mechanisms of merchant capitalism. During the eighteenth century the provisioning of Istanbul took place through a system of price and profit control. Even the volume of imported cereals was controlled, putting a hold on the powers that should tend towards the creation of a free market. The social system perpetuates its long-established economic equilibrium, and by so doing perpetuates itself.

However, the game unfolded on the edges of the Empire, where large centers of trade export were thriving, such as Alexandria, Izmir, and Thessaloniki. There, where the restrictive mechanisms were defused, particularly because of the effect of external demand which was intensified every now and then through the political and climatic conjuncture. Among the many examples include: the adverse climate conjuncture in the West from 1548 to 1564 that required the cereal production of the East; similar conjunctures in the eighteenth century overturn the usual relations between cereal influx in Marseilles in favor of the Eastern regions. Political conjuncture directs things towards the same direction: whether it is the treaty of Küçük Kaynarca, the continental blockade, or the effect of European warfare on Mediterranean transports, we encounter the same phenomenon, namely the strengthening of the Greek merchant navy that intensified its western routes. This is not merely the outcome of the positive conjuncture for the Greek merchant navy. Connections, if mediocre, are constant. Even when exporters prefer the Greek merchant navy, particularly locals, they do so for other reasons: more advantageous freight costs, the competition between the western powers, namely England and France in the eighteenth century. Yet, there were also other internal reasons.

These reasons concern capital strategy, which was particularly evident during the continental blockade. Ships were set into motion not necessarily because there was cargo, but because there were also monetary availabilities distributed between ship owners and the conduct of trade. The same people were simultaneously ship owners, both ship owners and captains, merchants, and financiers: they distributed their capital in ship shares, purchasing commodities, and shipping loans. In this way, the shipping industry developed its own economic logic and corresponding economic practice, consisting of a number of activities whereby one preconditioned the other.

Unequally distributed in time and space, it is very difficult, indeed impossible, to even indicatively trace the tonnage, number of vessels, and routes of the Greek merchant navy in the sixteenth and seventeenth centuries. This does not mean that there are no sporadic or dispersed references: 196 vessels are mentioned in Venetian sources from 1500 to 1544 (starting off from Crete in their overwhelming majority) with

Venice, Crete, Ionian islands, the Peloponnese, Syria, Istanbul, Aegean islands, Avlona, Ragusa, and Cyprus as their final destination. They demonstrate the inter-Mediterranean range of the journeys, with the sole exception a single trip to Flanders. However, this picture is not representative because it only refers to the Venetian network. As previously mentioned, Greek routes are mainly concentrated in the Eastern Mediterranean. Here are some more indicative facts. In 1764 there are 75 vessels in Messolongi, most of them built there or in nearby areas; overall tonnage was 10,640 tones and medium range tonnage 141.86. In 1813 there are 615 vessels in islands and ports, of general and middle range tonnage 153,580 and 249.72 respectively (obviously figures are exaggerated). At the end of the first 20 years, Hydra seems to have 186 ships of 27,716 overall tonnage. To continue these indicative cases regarding maritime transport, throughout the last 40 years of the eighteenth century, ships arriving to the free port of Ancona from the Ottoman Empire belonging to its subjects, Greeks presumably, did not exceed 5% of the overall arrivals. In fact, the numbers were slightly higher since the ships of the Ionian islands were considered Venetian. On the contrary, at the end of the century, Greek and Ottoman ships represented sixty to seventy percent of overall transports at the port of Alexandria. In Thessaloniki in 1819, out of 152 arrivals 100 were Greek, and twelve Ottoman; the rest were mainly divided between French and Austrians. In the same year, the share of Westerners in imports from the Western Mediterranean to Thessaloniki is 99.71 percent compared to 1.19 percent of Greek ships. Greek imports from the Eastern Mediterranean to Thessaloniki represented 98.81 percent of the overall Greek imports, Eastern and Western. Within the five-year period of 1815 to 1820, even though Greek presence in the port of Patras had the highest number of vessels, its predominance gradually decreased: in 1815 it represented 85% of overall tonnage, whereas by 1820 it only represented 45%.

These few examples illustrate the main, but not exclusive, focus of activity of the Greek merchant navy: the Eastern Mediterranean and the Black Sea. On the one hand, there were constant connections with the Gulf of Venice. On the other, there were wide-reaching connections that intensified with the French Revolution and the continental blockade; these were, nevertheless, established with fluctuations in the last thirty years of the eighteenth century. We will still have to include the transfers instigated by of Greeks living outside the Ottoman Empire. Some of them were ship owners, mostly merchants, residing in Mediterranean cities, South Russia, the Danubian Principalities, as well as the North. Their trade embraced all the whole variety of goods circulating in the Mediterranean. Alongside merchandize, they also import tastes and ways of life.

As we have seen, the relationship with the sea was not only peaceful. Constant wars, piracy, and the *corso* constituted an experience that in turn employed mechanisms of defense, both material and psychological. The relationship with the sea was also shaped through service in the Ottoman navy and Ottoman shipyards, as was the case with the Venetian network. Furthermore, the relation with the sea meant that Greeks participated in changes in naval technology. The westernization of Greek ships was one of the most distinguished examples of adjustment mechanisms. The same applied to the Ottoman navy, although its performance from the seventeenth century onwards when sail proliferated did not match the preceding one with oared galleys. In 1821 the role of Greek ships changed: from cereal-carrying cargo ships they became warships.

Greek ships had already been turning into military ones in order to counter Barbary pirates. By 1821 they were more or less ready to deal with an overpowering navy that was, however, less agile but also harder to be hit by their own range of fire. This navy had lost a large part of its driving force, namely Greek sailors. Taking advantage of the continental blockade, the latter had four naval bases: Hydra, first and foremost, Spetses, Psara, and Kasos. They had also accumulated money, which was most probably not destined for shipping. To counter the superiority of the Ottoman navy, they developed appropriate tactics. For this to happen though, they had to conquer certain necessary aspects of shipping organization: crew discipline, knowledge of the sea, internal hierarchy, and practice. All these features had already been acquired.

To counter the frontal attack of ships they employed concentrated attack. They capitalized on the slowness of big ships, using the agility of small ones. To counter the supremacy in fire power, they used an adjusted version of the fire-ship. Instead of drifting to its target through sea currents, it was now heading directly towards it, covered by the open fire of warships. This is, therefore, a readjustment of techniques and an optimization of potential. These could not have taken place without a previously-acquired knowledge, a supremacy on the sea through the experience of techniques of floating, so to speak, in a Mediterranean whose history is present with all its constraints, and at the same time provides the incentives for their transgression.

III
Insularity in the Greek World[4]

Like the rest of the Mediterranean, Greek islands do not have a unified character. Rather, they have mutual features, shared historical trajectories, or even common economic functions. Similarities, however, amount to neither unity nor centrality: they constitute a colorful multiplicity, often with a dispersed, yet often similar kind of chromaticism, which for this precise reason does not belong to the same tonality.

The islands of the Aegean and the Ionian seas encountered different and successive forms of rule during the early modern period up until most of them were gradually incorporated into the Greek state. As such, they were incorporated into cultural systems that transcended them, into systems of power that were imposed on them. Foreign rule also brought with it the establishment of internal systems of political and economic power, as well as the intermingling with different populations.

Religious dichotomies were not absent from these processes: Orthodox and Catholics, Christians and Muslims, and Jews. Although not every single island became a point of reception of such dichotomies, they nevertheless remained one of their general characteristics. The same applies to other kinds of cultural fusion, be they linguistic, artistic, architectural, or even juridical. If someone wanted to trace specificities alongside the mutually shared elements of islands, the fields of observation would be far from limited. They would expand beyond the differences of local economies and administrative systems, regional dialects or music, the spiritual hegemonies that are unequally expressed through local cults, dietary habits, the focus of intellectual phenomena, particularly educational institutions. Specificities are also found in the interior of islands, and not just between town and country.

Divided in two archipelagos, islands look towards a distant or adjacent land. Yet, apart from some rare exceptions, they do not constitute an administrative continuum with the mainland. The same cannot be said of economic continuums, i.e. the economic transactions of islands with lands across their shores. Climatic conditions determine the cultivation zones, which in turn mark the agricultural economy of islands. On the one hand, the distribution of rainfall regulates the extensive cultivation of olive trees in the Ionian and Aegean islands. On the other hand, aridity occupies a corresponding role for vines. In the islands of the two seas both crops are a shared feature of their economies, though of unequal distribution. Even though these are age-old crops, their relative position in the overall island economies changes over time. The seventeenth century is the era of the proliferation of olive and raisin cultivation: the former in the islands of both seas, and the latter in those of the Ionian. Both phenomena are the outcome of the linking of island economies with the broader and dominant northern and Mediterranean economies. But this connection does not exclusive rely on the demand for olive oil and raisins (or wine, in earlier times), and therefore does not only concern those islands where these commodities are produced. It is based on a broader and global circulation of goods, in which every island participates in its own way. Regardless of whether these circulations are directed outward or primarily inward, they eventually become a combination of the two. In other words, islands are not isolated, but they constitute a network of communications, which is weaker internally, but is nevertheless constant to the extent that these are not obstructed by weather or political conjuncture, namely war.

On the one hand, weather plays a role because the Aegean is not a quiet sea. Its winds are different and with local surges, interrupting transport for shorter or longer periods. In the case of the latter, they permit navigation from May to October. Navigation is not easy at all for those who do not know the sea, its reefs, and its safe havens. Those of the islanders who engage in maritime affairs have the insular nexus inscribed in their memory, perpetuating this experience.

On the other hand, war and the *corso*, as its collateral, paralyze peaceful voyages, interrupt the routes of trade, and condemn islands to isolation. Naturally, war is neither an endemic nor an exceptional condition.

Alongside its temporary transformations, it also causes permanent ones: changes in sovereignty, which in turn in foreign rule bring about cultural discontinuities. Such are the cases of Cyprus or Crete, and to a lesser extent Chios. Conversely, the change of political authority may also reinforce social and cultural continuities elsewhere by modifying the elements that compose them, as in the case of the Ionian islands.

One may argue that these are the general conditions that regulate the life of islands and islanders of the Aegean in the Ottoman temporal context, broadly speaking from the fifteenth to the eighteenth century. Nonetheless, islands have their own primordial characteristics: first and foremost, their own economy. Agriculture and animal herding are found everywhere. At the same time, there is also manufacturing and trade. The latter is in conjunction with the former two. However, in the context of exchange, trade acquires a relative autonomy not only at the level of transport, but also at the level of the composition of internal markets.

Large or small, with one or many settlements, islands utilize their arable and barren land. But land was not everywhere subjected to the same systems of use. Neither did units of cultivation have similar sizes. In Venetian possessions, such as Crete or Corfu, feudal institutions were transplanted, or earlier ones were assimilated—like those of the House of Anjou in the case of Corfu. Thus, territorial fiefs were created, in which cultivators were connected to the feudal lord with converging yet not identical relations. Elsewhere, small landholding took precedence without any other bonds of dependency than those prescribed by the Ottoman fiscal system. When the Ottoman conquest rested on an earlier system of land control, established by one or another western ruler, it retained some of its characteristics—for example in Andros or Naxos. Elsewhere, the Ottomans replaced it with their own land control system, maintaining however previous distributive systems: that is, the means of participation in the harvest involving both the lord of the land and the producer. In smaller islands, land was to a great extent fragmented. As units of measures of land surface were often derived from those of capacity, defining the size of land in cases of high fragmentation was done with small units of measure, also indicative of small seedings. Even if cereal cultivation was widespread, this did not presuppose each island's adequate grain production and eventual self-sufficiency in provisions.

Consequently, we are confronted with a discrepancy between the population potential and productive capacity of island societies.

This discrepancy did not necessitate an omnipresent, if relative, overpopulation in the context of the nexus of agricultural subsistence economies. It was also due to economic choices that resulted in the receding of cereal cultivation, losing ground to other commercially marketable crops. For example, during the first centuries of Venetian rule Crete was considered as a granary able to supply other Venetian territories like Euboea, a clearly exaggerated expectation. The policy of the ruler was based on securing cereal provisions, so that Crete would not have to "beg elsewhere" for its bread. During the sixteenth century, cereal cultivation receded to the benefit of vines: prices favored wine instead of wheat. Regarding grain provisions, Crete became subjected to imports from nearby ports. Since grain deteriorates quickly, the needs of the garrison and inhabitants of the cities would have to be met by baking hardtack biscuit. Furthermore, it was not only vines that sidelined cereals: in the seventeenth century, flax also contributed to their uprooting from the plain of Messara in Crete. The olive tree was another rival emerging from the second half of the sixteenth century. In this context, the paradigm of grain self-sufficiency, which rulers consider as a precondition of survival of their possessions, is destabilized. The very carriers of this shift were the feudal lordsmoving to Crete. Alongside them, agricultural producers participated in the market, or they were forced to participate, to the extent to which their own income made it possible.

What is observed here, then, is a turn to cultivations that responded to external demand. The latter was dictated by the needs of both manufacturing production and the dietary model of the societies and economies driving this demand: for example, Cretan wines as well as Ionian and Peloponnesian raisins. In both cases, the economic policy of Venice in its eastern possessions opposed this outcome. The same applied to the expansion of olive cultivation in Crete, the production of which was mainly intended for manufacturing rather than the food market. The proliferation of both olive and raisin cultivation went beyond the limits of Venetian possessions. They also included territories under Ottoman rule. This means that cash crops were incorporated in an economic system that transcended "state" control. If, for example, we compare what we know about olive cultivation during the fifteenth century with that of the seventeenth century, we will see that the quantitative data is diametrically opposed, whether in reference to the Peloponnese, Lesvos, or Western mainland Greece. This essentially means that during the late seventeenth century, olive cultivation did not respond to the needs of the local population, but to a broader demand mainly from manufacturing; this demand did not necessarily come from the West. Regarding the promotion of olive cultivation, Venetian policy records an exception in the case of the Ionian islands during the later seventeenth century. Seeking to reduce dependency from the oils of Apulia, Venice promoted olive tree plantations, particularly in Corfu. While this led to the imposition of a single-crop dependency on the economy of the so-called "eye of the Adriatic", i.e. Corfu, it does not alter the established relations of production: the promotion of olive cultivation was based on old planting systems, the same as those of vine cultivation. Cultivators were asked to plant or graft olive trees in order to become co-owners of the tree; the same was true of vines. In this context, the land-owners in this sharecropping relationship were usually feudal lords, alongside others who were outside the feudal network, or monasteries and churches. Thus, the response to the demand of a trans-local economy, whether this surpassed the network of the ruler or was indeed created by this network, did not necessitate a shift in the ways in which production was carried out. Rather, it contributed to reinforcing one of its factors, monetization. At the same time, it provided favorable conditions to local merchants and maritime transporters.

In brief, the shape of the agricultural character of the islands was formed during the seventeenth century. One of the main components of this landscape, the olive tree, would be further enhanced in the Ionian sea, Crete, and Lesvos during the following century. There were, naturally, other tree crops that were unequally distributed between and within islands: for instance, citrus or mastic trees in Chios. Receding cereals were substituted not only by vine and olive trees, but also by "industrial" cultivations, such as sugarcane and cotton in Cyprus—the latter during the sixteenth century. While cereals were a constant feature in the agricultural economy of all islands, production was not enough to achieve self-sufficiency, with the exception of some small islands. The typically Mediterranean polyculture that characterized island economies required specializations that exceeded the skills and experiences of a grain producer. Apart from transport, the mobility of goods entailed manufacturing activity that in turn required the import of raw goods: these included staves and hoops for barrels reaching Crete from Istanbul, Thessaloniki, or Venice. The same is true of agricultural tools.

Western rule, with Venice as the primary example, shaped the type of urban development in larger islands. Towns were fortified points, encompassing the mixed and stratified population. In Crete, westerners were on the top of the social pyramid: "Italian nobles" and feudal lords, to whom the indigenous *archontoromaioi* (Rum nobles) will be added. Such distinctions are even less noticeable in the Ionian islands. Moreover, in both cases nobility was accompanied by a local designation in order to distinguish local nobles—for instance, Cretan or Corfiot—from metropolitan ones. Only a rather small section of the urban population fulfilled the criteria for the class of *cittadini*. This social group held local power and authority that was not directly exercised by metropolitan officials. The body of *cittadini* occupied public offices, sharing among themselves the economic benefits that such positions entailed. In certain islands, such as Cythera or Cephalonia, they were also found outside of the urban center or the castle, residing in rural settlements. While this class did not constitute nobility for the metropolis, it effectively did so for the local societies, forming local aristocracies that were perpetuated until the incorporation of the islands into the modern Greek state. This, however, did not mean that they exercise an institutionalized and dominant role up to that point in time. New economic and social strata emerged, bringing about a redistribution of wealth: merchants, ship-owners, professionals, wealthy villagers, merchant landlords, men of finance who continue centuries-old usurious practices. Not only did the latter control the countryside, but also the very representatives of local aristocracies, especially during the nineteenth century. Next to the *cittadini* or aristocrats, there are the "plebeians" of the towns: artisans and professionals, the most powerful of whom managed to be incorporated into the body of *cittadini* through violent or peaceful means. What we, therefore, have here are towns with a social and economic stratification that was intensified by the existence of institutionalized social bodies.

These towns followed the western paradigm not only in their social organization, but also in the architecture of both public and private spaces. They were also western oriented regarding education. This is not only because noble families tended to send some of their children to study in Padua, or because Cretans, Cypriots, and Ionians wrote their works in Italian. More importantly, they used western literary production to shape their own Greek-language one, adapting the products of the former to their own cultural climate. By the nineteenth-century this cultural climate was already a national one in the Ionian Islands. Western-oriented intellectual phenomena are urban: when the Ottoman conquest of Crete altered the characteristics of towns, dispersing their leading strata, Cretan literature would disappear from its place of birth. The same had already happened in Cyprus with regard to its cultural production. This continued mainly in printed form as a shared part of Greek culture. From an early stage, island consciousness found its expression mainly through the histories of islands (Cyprus, Crete, but also Zakynthos), usually in Italian but also in rarer instances in Latin as well as Greek, with references to antiquity. These are "local" histories: each island is examined separately rather than all together. For instance, in order to acquire their historiographical unity, the Ionian Islands first needed to have formed a state in the early nineteenth century.

Larger islands were characterized by deep dichotomies between town and countryside: the feudal lords of the latter had, or were forced to have, links with the town, or indeed to reside there. Their revenues were transferred to the town both as natural or, mainly, exchange values. In other words, the rural surplus was absorbed by the town. In addition, the countryside was indebted to the town, and in this way the latter exerted further control over the rural surplus. This was because the composition of crops did not cover the subsistence needs of agricultural households, which were therefore dependent on the market for their nutrition. The stereotypical image of the cultivators of feudal lands in Crete, in sharp contrast to that of mountain populations, was one of a scared person whose body language reflected fear and submission. As previously mentioned, the turn to cash crops led to the diffusion of monetization in the countryside. This, however, did not mean that monetary transactions were a generalized phenomenon. Money functioned as a loan repaid in kind (natural values) at the time of harvest, usually calculated on the basis of low prices due to seasonal compression. In smaller islands, where cash-crops were less prominent, monetization comes as the result of maritime and trading activities. The latter also drove local cottage industries or manufacturing goods: textiles, socks, straw hats, hand-mills or other stone tools; there were also sponges in the cases of those islands specializing in sponge fishing. The manufacturing of hand-woven and textile goods was widespread. The former particularly thrived in Chios, where a large quantity of silk was used; this was locally-produced as well as imported. The making of the thread and the weaving took place at home, based on orders from the merchants trading these goods. In other words, they were not produced by manufacturing workshops for the market. Traveling merchants also placed such orders, adding to the main volume of their goods small quantities of the products of cottage industry, which was usually a female affair.

The Aegean islands were much more connected with multiple maritime transport links than those of the Ionian. Aside from the traffic of naval fleets, the mobility of commercial ships caused different kinds of connections, in which islands had their share, shaping through them their economic geography. There was also the large volume of traffic of northern and western ships, connecting the ports of the Eastern with those of the Western Mediterranean (mainly Venice and Marseilles), as well as with the cities of the North. Such connections concerned few islands, mainly those whose production interested western and northern markets: Cyprus, Crete, and most of the Ionian Islands. One could add to this list islands that were connected through this kind of transport, for example Chios or Lesvos. French ships served a supplementary route from island to island, the so-called "caravan," through which an important part of trade was conducted. Parallel to these types of transport, there were also others, which were served by local fleets linking the islands between them and with the mainland coasts. This does not mean that local fleets were limited to these routes. From an early stage, in the sixteenth century, they participated in large-scale maritime traffic in those routes that were part of the Venetian network. Local fleets dominated the maritime transport of the Eastern Mediterranean, and in the nineteenth century that of the Black Sea and the Danube. Nonetheless, voyages to the ports of the Adriatic were constant, especially by Ionian ships in the eighteenth century. Historical conjuncture created such opportunities so that the fleets of particular islands played a special role—as was the case of Hydra during the years of the Continental blockade. Finally, not all islands had fleets, but all were served by the fleets of those islands that do.

If some island ships served their local exports, other ships, like the Hydriot ones, were part of a broader nexus of commodities transport that did not belong to their own economy. Such was the case of the trade of southern Russian cereals. Nevertheless, all ships returned to their home base once their commercial voyage was completed. They did so to clear their accounts and to bring goods for the local market, for clients, or for the homes of sailors. Temporary companies were formed in the places where the ships originated from, either for the export of particular local goods and the import of others, or in search of cargo in other ports. Consequently, maritime transport mobilized commercial or financial interests in the places of origin. Such interests may have been average or important ones; the latter appear where wealth was concentrated, Hydra being the exemplary case. Depending on the degree to which those involved in a maritime trading voyage are engaged, they all became active contributors to and participants of a maritime enterprise, internalizing its economic logic. Thus, ship-owners and sea captains were not just transporters who limited themselves to collecting a freight fare; they were concurrently investors of a capital that would increase either as a maritime loan or as commercial capital. Sailors may have had a share in profits instead of a salary. They may also have had a right in the *pacotiglia*, the free carriage of goods in exchange for split profits between the merchant and the captain. In this fashion, they became active partners of a commercial enterprise. These climaxes in the islands' economic functions are naturally neither generalizable nor do they have a stable presence over time, even if they can be detected sporadically. Furthermore, they do not concern all island populations, or indeed those populations dealing with agricultural economy. In the latter, particularly so in large islands, the relationship with the sea was almost hostile, for the sea translates to forced labor in the galleys. Nonetheless, to the degree that their economy produced an exportable surplus, they were still subjected to the sea, even if only indirectly.

We tend to study trade from the vantage point of large-scale phenomena. Yet, these are divided to infinite pieces, which in turn affect people's everyday lives. Small island communities testify to the importance of the diffusion of commodities that come from an economy different to their own. This importance was both economic and cultural. Putting the large-scale phenomena aside, let us attempt to examine what was being imported into the islands. Where subsistence was not guaranteed and its supplementation is dependent upon maritime and commercial transport, we encounter a multitude of goods coming from the markets of the West. These goods were incorporated into the lived experience of island societies, performing multiple roles, whether these are purely functional, or related to hoarding or the acquisition of prestige. These commodities vary: from fishing hooks and nails to textiles (considered as luxury products), and from silverware to gold jewelry. The list could go on. In any case, it would mark a process of cultural fusion that was not the prerogative of the dominant social strata: these were popularized goods that indicate the mechanisms of a harmonious civilizational process, the shaping of tastes and ways of life.

Both the Aegean and Ionian islands were acentric, in the sense that none of them has performed the kind of economic roles that would subjugate to them all other islands. The Ionian Islands experienced forms of administration that created a fragile unity; this was enhanced after the end of Venetian rule, when they were called to participate in a central administrative body, i.e. a parliament and a government. Yet, their dependency to the mainland across the islands was never abolished. Not only was the latter a source of grain provisioning, it also gave seasonal work to their local labor, i.e. the land workers that could not support their subsistence in their places of origin. Nonetheless, these acentric islands form networks that connect one island with another, and most of them with the immediate mainland, both to their East and their West. They also constituted local communal structures that perpetuated judicial and economic cohesions, as well as the solidarities imposed by the Ottoman fiscal system. The economic network of the islands was individualized in internal communications, finding its matching rhythms in a cultural network that also lacked a unified form; rather, it is best described as mutually shared elements that affected all cultural fields and behaviors. This cultural network concurrently expressed and defined insular polycentrism in varying densities and intensities.

CONCEPTUALIZING INSULARITY

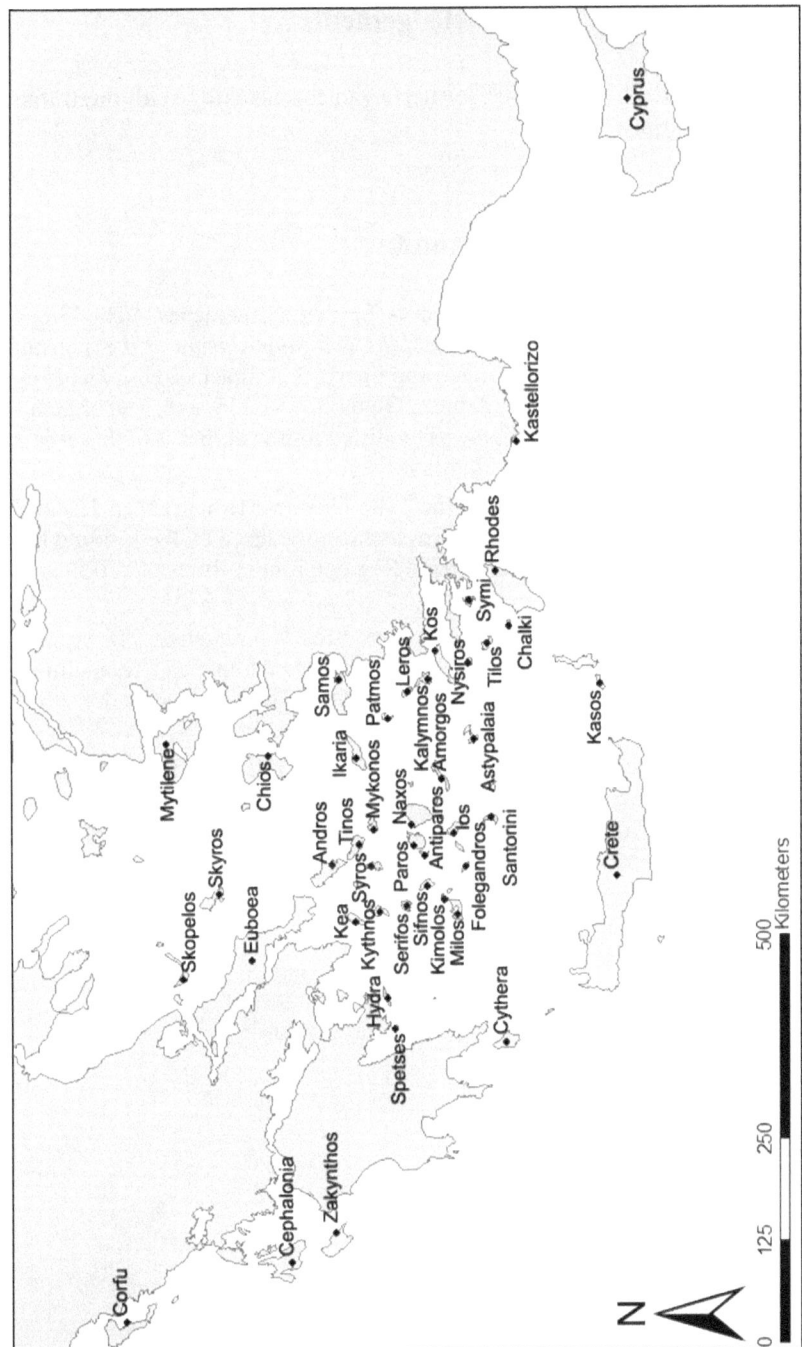

Map of the islands mentioned in the text

Acknowledgements

The editor would like to thank Eleftheria Zei for assisting with the translation of certain terms.

Notes

1. The essay was originally published as Spyros I. Asdrachas, "The Greek Archipelago: A Far Flung City," in *Maps and Map-Makers of the Aegean*, Vasilis Sphyroeras, Anna Avramea, and Spyros I. Asdrachas, eds., Geoffrey Cox and John Solman, trans. (Athens: Olkos, 1985), 235–248. The originally published English translation was modified for the purposes of this article by Anna Papaeti and Antonis Hadjikyriacou.
2. Unpublished paper presented at the 'The City in Mediterranean History' session of the International Study Conference organised by the Instituto per i beni artistici, culturali e naturali della Regione Emilia-Romana in Bologna, November 1983.
3. This essay was originally published as Spyros I. Asdrachas, "Τα νησιά," [The Islands] in *Οικονομία και Νοοτροπίες* [Economy and Mentalities] (Athens: Ermis, 1988), 237–244. Translation from the Greek by Anna Papaeti.
4. This essay was originally published as an afterword in Giorgos Tolias, *Τα νησολόγια: η μοναξιά και συντροφιά των νησιών* [Isolarii: the Solitude and Companionship of Islands] (Athens: Olkos, 2002), 169–176. Translation from the Greek by Antonis Hadjikyriacou.

The Historiography of Aegean Insularity

ELEFTHERIA ZEI

The Aegean space has retained the attention of contemporary international historiography mainly because it seems to fit perfectly the two opposed trends regarding insular spaces in contemporary approaches: one stressing the character of "isolation" and extreme locality of micro-insularity, and the other, the "polynesian" character of insular States, which have been known, especially since the 1982 Jamaica International Convention, as "Archipelago" States.[1] Although in touch with the above international trends, historiography on Modern Greek history did not but very recently participate in the debates they raised. Instead, it opted for the annexation of modern Aegean space to its more urgent internal problematic: the capitalistic formation of nineteenth-century revolutionary and post-revolutionary Greek society and the Greek State.

1. "Archipelagic" Models versus the "Isolation" of Small Islands

Modern geographical, historical, and anthropological literature on insularity has generally invested in a dichotomy: on one hand, the theme of "isolation," a very ancient concept of insularity traced back to the Latin roots of the word "island" (*insula*, *isola*), which was rehabilitated by modern geographic and ethnographic theories; and, on the other hand, the theme of interconnected insularity, brought forward by contemporary Mediterranean theories. Resuming the geographical traditions of Friedriech Ratzel and Paul Vidal de La Blache, and bearing strongly the echo of English naturalistic and anthropological evolutionist theories (A.R. Wallace),[2] late-nineteenth and early-twentieth-century German and French human geography focused on the "isolation" and "particularity" of insular species and cultures.[3] Any larger insular complexes were seen as rich fields of research for single island worlds: thus, for instance, the study of the Mediterranean Sea by Ratzel or the Malay Archipelago by Wallace, follow the same line of thought as the "isolate" models in contemporary biological and population studies.[4]

A different "archipelagic" insularity has been largely explored in more recent Mediterranean theories, mainly re-discussing the Braudelian geographical, historical, and human vision of the Mediterranean. In 2000 the Braudelian model was further elaborated by Peregrine Horden and Nicolas Purcell's Mediterranean,[5] presented as a geographic, social, and cultural area of multiple distinct *localities*. Among them, islands play an important role, characterized by different possibilities of interaction and connectivity; later, the Braudelian vision was contested by David Abulafia's global conceptualization of the Mediterranean as a sea of converging differences.[6] Despite their oppositions, all the above comprehensive archipelagic models assimilated single islands in a succession of physical and conceptual complexes or networks, or as Patrice Brun would have put it, as "Archipelagoes in the Archipelago,"[7] with the common intention to overcome the isolation issue. Nevertheless, all connectivity approaches, encompassing further distinctions of insularity invented outside the Mediterranean archipelagic model, such as "oceanic" or "continental" islands in more recent island theories,[8] seem to have created a new problem of isolation; this is due to the tendency to leave out of their field of studies remote, sparsely populated, or uninhabited isles and islets.[9] In consequence, at least up until the beginning of the twenty-first century, classical studies—mainly archaeology and ancient history—have explored the "micro-insularity problem" as both a problem of scale and connectivity,[10] seeking for case studies in the Indian Ocean, the Polynesia, the Malay or the Japanese archipelagoes, as well as in the Aegean Archipelago. The analytical category of micro-insularity thus established was very much related to the "Island metaphor" of anthropologists,[11] or even to nineteenth-century "population isolates." The impasses of the above dichotomy as well as the extension of island studies to other cultural eras and areas (for instance, the Ottoman Empire), led scholars to re-open the debate, and explore different forms, uses, and representations of *insularity* beyond connectivity:[12] thus, for example, micro-insularity offered a privileged field of research for the most recent approaches and debates in frontier-border studies.[13]

Although Lucien Febvre's introduction and development of the notion of "insularity" in historical geography discussed the determinism of geographical "isolation," it was only in the 1970s, and under the influence of Fernand Braudel and his vision of the Mediterranean as a geographical, historical and cultural entity, that a decisive global work on the Greek Archipelago saw the light: Émile Kolodny's three-volume thesis of historical geography and demography, *La Population des îles de la Grèce*.[14] Summing up older and modern approaches, Kolodny's work contested the theme of natural geographic "isolation" by placing Greek insular complexes inside Mediterranean insularity. The Aegean islands offered a particular field of "insular geography," not only by the excessive division of surfaces and settlements, but also by the development of historical itineraries that were independent from the continent. Kolodny, thus, introduced the idea of a historically generated economic and social isolation of the islands, which he attributed to the continental organization of the Greek State in the nineteenth century, and the subsequent economic and administrative subordination of the islands to the coast.

Around the same period, in Braudel's close environment, the Italian economic historian Ruggiero Romano suggested that the Aegean insular complex could be regarded as a single economic space, a Mediterranean modern "urban market."[15] Its mechanisms would be first approached by Spyros I. Asdrachas in his unpublished doctoral thesis (1972),[16] to be further developed in his article on the Greek Archipelago as a "dispersed city" (1985).[17] In 1992, elaborating on the Wallerstein model, Traian Stoianovitch would describe the Aegean "market" as a peripheral system of periodically recurring crises, "of cyclic instability," tributary to a larger Mediterranean-Balkan economy.[18] In a renovated "center-periphery" interpretation, it is important to note that in his work *L'Économie agricole grecque*, Socrates Petmezas indirectly suggests two lines of research, which still remain unexplored: the relation of the Aegean market "periphery" to the economic "center" of nineteenth-century Greek state, and the possibility of its relations with North-Western or Central European markets.[19]

In light of network theories, recent critical approaches of the Aegean insularity and the archipelagic model have also been attempted through Braudel's Mediterranean concept,[20] as well as through Romano's and Asdrachas' Aegean model.[21] Apart from economic history, Maroula Synarelli's research at the Laboratory of Historical Demography (École des hautes études en sciences sociales - EHESS, Paris) presents an interesting attempt at reconstructing a seventeenth-century Aegean model of mobility networks through individual and family-life itineraries during an exclusively transitional period between Latin and Ottoman installations.[22] The collective volume published on the occasion of the 10th International Exposition in the Biennale of Venice in 2006, endeavors the introduction of the above model in contemporary architectural and urban studies of the Aegean space.[23] Finally, network models have been more recently elaborated in the study of the Greek Archipelago as a particular modern "topos" for the establishment, maintenance, and extension of maritime commercial and warfare networks.[24]

2. A Long-Term Supremacy of the Local: the "Community" Theme in the National Historiography of the 1980s

Nevertheless, the above conceptualisations of the Greek Archipelago as a united demographical, economic, and social space, as well as its Mediterranean affinities left little imprint on the Greek historiography before the 1980s. Afterwards, and in the political climate following the end of the dictatorship of the colonels, Greek historians turned to extended network of complexes of islands and islets of the Aegean insular space as an ideal case-study for "Modern Greek History";[25] that is, the history of Greek territories under Ottoman domination (fifteenth to nineteenth century) and the transition to the Greek State (nineteenth to twentieth century), through newly introduced domains, such as economic history, historical geography, and historical demography. A number of works on the Aegean islands were then conceived or saw the light through the graduate program on Post-Byzantine Studies (Paris I) under the direction of Spyros Asdrachas; they were also supported by research programs on rural history and historical demography at the National Hellenic Research Foundation (NHRF) under the direction of Asdrachas and Vassilis Panayotopoulos.

Most of them bore the character of case studies in economic history from the eighteenth-century Aegean. Reproducing a French tradition of the historical *exemple*, these studies worked with basic quantitative methods and followed two distinct guiding lines. The first drew on a classic marxist problematic whereby the above case studies were designed to contribute to the historical construction of the economic profile of Greek insular populations, and their integration in, or resistance against the Ottoman economic/fiscal system in view of the formation of a Greek capitalist society.[26] The traditionally commercial profile of the islands was, therefore, studied in this light. The observations of Vassilis Kremmydas on insular commercial structures and practices fell in step with the classic model of the capitalist turn in Greek maritime and commercial his-tory at the end of the eighteenth century, arguing for the supremacy of the islands of the Saronicos Gulf.[27] Along the same lines, Asdrachas' doctoral thesis dealing with the formation of commercial capital in the islands, Asdrachas extended the same problematique to the development of eighteenth-century economic mentalities and culture in the islands.[28]

The second guiding line was a long-debated theme in Greek historiography of "Greek communities" under Ottoman rule. This was revived by Asdrachas' pioneering articles in the *Istorika/Historica* journal, in which he studied the role of fiscal functions in the economic and social transformation of eighteenth-century insular communities and their notables.[29]

Revolving around the "community-fiscal axis," the first to appear in chronological order was Eftychia Liata's *Seriphos*,[30] followed by the unpublished thesis of Sevasti Lazaris on Mykonos:[31] both dealt with the Greek tax cadasters of the above islands, venturing to identify an economic profile of insular society through the redistribution of tax and land property inside the insular "community." Through the economic profile and roles of the communal authorities of Mykonos, Lazaris' thesis in particular suggests another line of study: the history of island elites, a topic to be further discussed below. Along the same lines, Evangelia Balta and

Maria Spiliotopoulou published an article in 1996 on Santorini,[32] inaugurating the promising combination of research in both Ottoman and Greek cadasters. The article also announced another field of research, that of individual and family economic activities and practices in the Ottoman Empire, bringing forward the land and agricultural aspects of insular economy on the basis of local notary acts. This line of research was explored by Dimitris Dimitropoulos' doctoral thesis on Mykonos,[33] and later by Spiliotopoulou's doctoral thesis on Santorini.[34] Mainly due to reasons of documentation, the above case studies referred mostly, but not always, to the insular complex of Cyclades, where research often referred to the central settlement of the island. The archival material used consisted of tax cadasters in the Greek language, and notary acts found primarily in the National Archives, as well as in local monastic and municipal archives.

The above production represented an important contribution towards a new modern history of the islands and Aegean space, as it contested several aspects of the traditional nationalist approaches of the nineteenth and the first half of the twentieth century. These included the exclusively maritime character of insular societies, the romantic idea of insular poverty and precarious life, and an intrinsic model of resistance during the Greek Revolution, implying some sort of exceptionalism. They offered a more accurate, internal view of the islands in more than one aspect of their economic activities and role in the Ottoman Empire. Through their affinities with French historiography, the turn to Mediterranean studies, and the introduction of archipelagic models, the above doctoral theses endeavored to reconstruct a global insular space, whose existence was otherwise suggested mostly through commercial relations between the islands in question and the Archipelago, or the Mediterranean.[35]

The above historiography, although contesting the legal-administrative representation of the semi-autonomous Greek "community" under Ottoman rule embedded in traditional Greek historiography during the 1960s, did not altogether cease to study insular "communities" as consensual social spaces, characterized by the solidarity of their members against the Ottoman State.[36] A decisive turn in European historical interest that saw the Aegean insular space as a privileged area of modern social transformation was initiated in the beginning of the 1980s by

B.J. Slot's two-volume work entitled *Archipelagus Turbatus*.[37] The author studies the Cyclades as an archipelago caught between Latin "colonization" and Ottoman "occupation." He investigates the administrative, economic, and social transformations of the islands between a western and an eastern rule through Greek, Western, and Ottoman sources. It was not until the end of the 1990s, though, that questions of British "social history" made their appearance in historical literature on the Aegean islands. The publication of Aglaia Kasdagli's doctoral thesis, *Land and Marriage Settlements in the Aegean* (1999),[38] contesting Slot's theory about the eastern (Byzantine) origins of administrative structures in the Aegean, studies the norms and transgressions of property distribution in the feudal context of seventeenth-century Naxos through women's dowry and marriage arrangements.

Recently, new social aspects of the Aegean islands entered Greek historical inquiry under the influence of other disciplines such as social anthropology. During the 1980s and 90s social questions were dealt with through the economic and demographic approaches of insular history. The exchange between history and anthropology in the domain of modern studies of the Aegean was mainly represented by a few case studies, in the context of graduate courses in ethnology and social anthropology in the EHESS taught by Fritz Staal and Maurice Godelier.[39] These examined the impact of parental succession systems on the development of the built environment and inhabited space, on the social organization of ethnic settlements, or on the social impact of religion through specific local rituals.[40]

3. Historical Geography and Demography: An Attempt towards the Global

Between the 1980s and 90s, Vassilis Panayotopoulos initiated an ambitious research project on Greek historical geography and demography at the NHRF, focusing on the cartography and typology of Modern Greek settlements. Based on a particularly rich corpus of material, and following a French "laboratory" model, a new line of global approach of the Aegean space was ventured, which aspired to overcome the case-study model. Characterized by meticulous research and classification of population numbers and sites, project participants seem to remain conscious of the limitations of their historical material as well of research itself; in particular, for the larger part historical material were non demographical, consisting mostly of Greek fiscal sources until the beginning of the nineteenth century and the first population census.[41] However, the publications produced in the context of the above NHRF project present a systematic global survey of two important contributions to contemporary and future research: (a) insular toponymy, a particularly obscure field of research in the islands; and (b) insular cartography, based on cartographical as well as textual sources, the latter in collaboration with other NHRF research projects.[42] Overall, despite a prolific production and collection of case studies,[43] the rest of the works on Aegean historical demography generally remained quantitative and descriptive in their analysis. Conversely, works on the demography of the Ionian islands have been more successful in introducing new analytical tools and techniques, venturing to construct interpretative models for population movement and mobility in the Ionian area of the Mediterranean.[44]

4. Emphasis on Nineteenth-Century Transformations

During the 1990s Greek historiography placed the Aegean islands in the Eastern Mediterranean periphery, mainly in order to study them as separate fields of resistance or integration in the industrial and capitalist transformation of the nineteenth century. The study of nineteenth-century insular settlements and ports through combined historical domains and analytical tools, such as industrial archaeology, history of technology, network theories, and family history, showed new possibilities in the hitherto cultivated trends of economic history and commercial history. Questions on the history of technology in island agricultural production and maritime activities,[45] or ship-building,[46] permitted the study of the industrial transition of insular economies and societies.[47] It is important, though, to note that these new approaches eventually "annexed" Aegean space to continental history by employing the tools of broader fields of historical research such as urban history and industrial archeology. These fields served more a renewed interest in the study of state formation and industrial transformation in the 1990s, at the expense of an internal history of insular technological structures and economic changes in early modern times.

The example of the city and port of Hermoupolis has been particularly explored since the 1980s; its commercial and industrial role in the Eastern Mediterranean was examined in Vassilis Kardassis' *Syros*.[48] Since the 1990s, Hermoupolis has offered an interesting field of research to more recent sub-fields of history, such as urban history and industrial archaeology. Research on the island of Syros, conducted by the NHRF and the International Committee for the Conservation of the Industrial Heritage (TICCIH) was presented in a collective volume published in 2008.[49] Christos Loukos' work on Syros merits particular attention, as the author deals with a quite different aspect of the insular settlement, exploring the relationship between attitudes towards death and the formation of an insular ninenteenth-century bourgeoisie.[50]

Since the 1990s, there is a proliferation of studies on Greek commercial diaspora, its settlement and entrepreneurial activities in the Ottoman Empire and the Greek State from the eighteenth century onwardsand the participation of Aegean merchants, ship-owners, and entrepreneurs. Indicative are Gelina Harlaftis' *A History of Greek Owned Shipping*,[51] and the collective volume *Following the Nereids*,[52] a volume that includes a selection of papers from the Fourth International Congress of Maritime History (Corfu, 2004). Although referring to network theories, the above works mainly insist on tracing geographical individual and family networks of few famous insular cases, such as Chios and the islands of Saronicos Gulf, e.g. Hydra with the continent. However, they fail to produce an interpretative model of the Archipelago entrepreneurial activities, whose human networks still remain less explored.[53]

Earlier work by Nikos Belavilas, published in 1997, explored the connection between insular ports, fortifications, and early modern pirate commercial networks.[54] It offers not only a model of Aegean insular commercial networking and practices, but also a much more accurate, rather inverted picture from that of traditional Greek historiography on piracy.[55] In this fashion, Belavilas' work pushes further the debate on the role of piracy in the development of insular settlements.

5. Interdisciplinary Reflections and New Debates

The number and diversity of case studies produced, the voluminous documentary material revealed and explored, the opening to Ottoman history and archives, as well as the adoption of non-historical approaches and conceptual tools has recently led to the elaboration of a more refined, complex and interdisciplinary problematique on Aegean insular space. New tendencies were already announced in Asdrachas' short articles on islands. Originally published in the *Kathimerini* newspaper between 1992 and 1994, they were collected in a volume in 1995.[56] They were later summarized in the two-volume publication of the works of the Second European Congress on the Greek islands,[57] and in another collective volume of 2004.[58] The latter introduced more recent approaches, among which a history of the Aegean piracy in the context of Mediterranean corsair networks, an analysis of literary texts, of popular discourse and popular music, of visual representations, or a nocturnal history of the Aegean. In 2014 a new collection of newspaper articles by Asdrachas, brings forward, among other things, further intellectual approaches of Greek "micro-insularity" since the early modern period.[59] The notion of space is a crucial topic traditionally registered in ethnographical and architectural approaches. In her doctoral thesis (Athens 2002), Evdokia Olympitou (Ionian University) endeavors a first connection between insular population, space, society and production, which goes further than earlier architectural approaches of insular settlements.[60] Aegean insular spaces have also constituted the subject of a recent inquiry on internal and external borders. In the context of a workshop held in EHESS (Paris) and NHRF (Athens) in 2001–02 entitled *Histoire des frontières. Jeux de construction*, whose proceedings were published in 2008, the Aegean islands were considered in light of the geographical and socio-political frontier and its political-ideological exploitation during the nineteenth and the twentieth centuries.[61]

The Greek translation in 2001 of the French sociologist Bernard Vernier's work on parental systems of succession in Karpathos,[62] marked a decisive cross-point between history, sociology, and anthropology in the study of the islands. According to Verniers, parental systems of succession were no longer regarded as the legal norms of insular communities, but as institutionalized succession practices which ensured reproduction of elites, not only on the economic-social, but also on the symbolic level. This was made possible through the transmission of land, objects, and names, as well as through social construction of gender identities; this line of inquiry had already been inaugurated in 1991 by the collected volume *Contested Identities: Gender and Kinship in Modern Greece*.[63] A seminar organized in Hermoupolis (Syros) in 2004 by the NHRF brought together historians and anthropologists, different lectures and interpretations of insular archival material, mainly notary archives, Greek and Ottoman cadasters, and a diversified agenda of themes (servitude, community and church, marriage and death, concepts of time), in an effort to formulate a combined approach of the Aegean societies. Despite the richness of approaches, the seminar revealed the impasses of an interdisciplinary dialogue between the socio-economic and the anthropological conception of social changes and identities. Critical exchange with the above conceptualizations as well as with recent anthropological-cultural approaches of violence, social and gender identity in the insular society (such as that of ritualized male violence and its integration in nineteenth-century *Rechtsstaat*),[64] led to new lectures and to a new inquiry on the role of women and on relations of power in the islands;[65] these were traditionally considered more "revolutionary" (the Saronicos islands, Psara, Mykonos) during the 1821 revolution.[66]

Finally, exploring recent trends in elite theories (for instance, political and cultural "mediation") in state history and discourse analysis, historians have formulated, through the study of administrative, political and social interactions between institutions and discourses, a more complex understanding of the transformations of insular societies in modern times and under different forms of administrative or political subordination.[67] In a pioneering collective volume published in 2004, Gilles Veinstein calls for working on more than one "insularity" inside the Ottoman Empire.[68] The case of Paros illustrates the transformation of economic and social structures from Latin to Ottoman administration in the formation and reproduction of insular "notability" and its culture.[69] It opens the discussion about mountain structures in the Aegean, hitherto studied only in the case of large islands (Crete) and in their Mediterranean context. On a similar line and based on Ottoman archives, including Ottoman judicial sources, Elias Kolovos studies the social development of Andros, focusing on a new local elite of notables through the installation and transformation of Ottoman fiscal and judicial institutions from the sixteenth to the nineteenth century.[70] In fact, the study of insular fiscal and judicial institutions is a new field of research in international historiography. Early modern historians have also started to explore forms of social transformation and representation of Cycladic nobilities in the still scarcely studied political frame of the Duchy of Naxos.[71]

In conclusion, international historiography on insular spaces has turned its attention to the Aegean case since the beginning of the twentieth century, discussing insular complexes through the dichotomy isolation-connectivity and beyond. Nevertheless, twentieth-century Greek historiography on the Aegean islands in modern times did not join the above debates until late in the century. Instead, it opted for studying the modern Aegean space as a particular part of Modern Greek history, focusing on the role of the islands in the transformation of Modern Greek society since the eighteenth century.

Recent international geopolitical debates around "insular states" and their newly re-discovered historical "topos" (such as the Greek Archipelago, the Scandinavian archipelagoes, or the Japanese Archipelago) seem to have raised new questions, embracing and enhancing the Aegean paradigm on an international level. The above debates are currently considerably enriched by contemporary scientific research in Greece, with the development of interdisciplinary fields of study (combining mainly history, social anthropology, archaeology, architecture and urbanism), and the organization of international projects.[72]

Notes

1. The term "archipelago" first appears in the thirteenth century Venetian administrative language designating the Aegean area of the Republic's *Stato da Mar*; in the sixteenth century, it was officially integrated in the European dictionary, generally indicating a "sea of islands."
2. Alfred Russel Wallace, *Island Life: The Phenomena and Causes of Insular Faunas and Floras* (London: Macmillan, 1880).
3. Emmanuel de Martonne, *Traité de géographie physique* vols 1-2 (Paris:

Armand Colin, 1909); Jean Brunhes, *La Géographie humaine* (Paris: Félix Alean, 1910); Edgar Aubert de la Rüe, *L' Homme et les îles* (Paris: Gallimard, 1935); and Paul V. La Blache, "Les Particularités géographiques des îles," *Pages Géographiques* (1963), 187–208.

4. For instance Alfred Sauvy, *Théorie générale de la population*, vol. 1 (Paris: Presses Universitaires, 1963), 42–56.
5. Peregrine Horden and Nicholas Purcell, *The Corrupting Sea: A Study of Mediterranean History* (London: Blackwell Publishers, 2000).
6. David Abulafia, *The Great Sea. A Human History of the Mediterranean* (London: Allen Lane, 2011).
7. Patrice Brun, *Les Archipels égéens dans l'Antiquité Grecque (Ve-IIe siècle avant notre ère)* (Paris: Les Belles Lettres, 1996); see also Christie Constantakopoulou, *The Dance of the Islands: Insularity, Networks, the Athenian Empire, and the Aegean World* (Oxford: Oxford University Press, 2007).
8. For a general contemporary typology of insularities, see Roxani E. Margariti, "An Ocean of Islands: Islands, Insularity and Historiography of the Indian Ocean," in *The Sea: Thalassography and Historiography*, ed. Peter N. Miller (Ann Arbor: University of Michigan Press, 2013), 199–229.
9. Included in an ambitious review of further physical, cultural or intellectual distinctions, in Marc Shell, *Islandology: Geography, Rhetoric, Politics* (Stanford: Stanford University Press, 2014).
10. Patrice Brun, "Problèmes de la micro-insularité, en Grèce Égéenne: les exemples de Folégandros et de Sikinos," *Revue des Études Anciennes* 98, no. 3 (1996): 295–310.
11. Marshall Sahlins, *Islands of History* (Chicago: Chicago University Press, 1987); Thomas Hylland Eriksen, "In What Sense do Cultural Islands Exist," *Social Anthropology* 1, no. 1 (1993): 133–147; Robert C. Kiste and Mac Marshall, *American Anthropology in Micronesia: An Assessment* (Honolulu: University of Hawaii Press, 1999).
12. It is interesting to note the variety of terms in use: *island studies, nissology, thalassography, islandology.*
13. Michel Bertrand and Natividad Planas ed., *Les societies de la frontière de la Méditerranée à l' Atlantique (XVIe-XVIIIe siècle)* (Madrid: Casa de Velasquez, 2011); Linda T. Darling, "The Mediterranean as a Borderland, *Review of Middle East Studies* 46, no. 1 (2012): 54–63. See also the topics discussed in the recent international conference entitled "Insularities Connected: Bridging Seascapes from the Mediterranean to the Indian Ocean and Beyond," organized in Rethymno on the June 10–12, 2016 by the Institute of Mediterranean Studies (IMS)/Foundation for Research and Technology (Crete) and the York University of Canada. François Taglioni

goes even to the extent of disputing the very field of island studies, in a paper given in 2012 in Madère, entitled "Revue conceptuelle et critique de l'insularité. La nissologie est-elle une science?".

14. Émile Colodny, *La Population des îles de la Grèce. Essai de géographie insulaire en Méditerranée orientale*, vols 1–3 (Aix-en-Provence: Edisud, 1974).
15. Ruggiero Romano, "The Watery City of the Aegean," in *The Dispersed Urbanity of the Aegean Archipelago. 10th International Exhibition of Architecture Venice Biennale, Cities, Architecture and Society Director: Richard Burdett. Greek Participation,* eds. Elias Constantopoulos, Korina Filoxenidou, Katerina Kotzia, and Lois Papadopoulos (Athens: Hellenic Republic, Ministry of Culture, 2006), 35.
16. Spyros I. Asdrachas, "Patmos entre l' Adriatique et la Méditerrannée orientale pendant la deuxième moitié du XVIIIe siècle d'après les registres de Pothitos Xenos" (Paris: unpublished PhD. Diss., EHESS, 1972).
17. Spyros I. Asdrachas, "Το ελληνικό αρχιπέλαγος, μια διάσπαρτη πόλη," *Χάρτες και χαρτογράφοι του Αιγαίου* [Maps and Cartographers of the Aegean] (Athens: Olkos, 1985), 235–248.
18. Traian Stoianovitch, "Pour un modèle du commerce du Levant: économie concurrencielle et économie de bazar 1500-1800," *Between East and West. The Balkan and Mediterranean Worlds, 3. Land, Lords, States and Middlemen* (New Rochelle, N.Y.: A.D. Caratzas, 1992), 39–88. In 1996 the social urbanist Pierre Veltz introduced a model of "archipelagic economy," an intercontinental economic system based on a network of centers represented by the world metropolis. See his *Mondialisation, villes et territories: une économie d'archipel* (Paris: PUF, 1996).
19. Socrates Petmezas, *L' Économie agricole grecque. La dimension périphérique* (Heraklion: Éditions universitaires de Crête, 2003).
20. See for example Alexandra Sfoini ed., *Καλοκαιρινές Σαμιακές Συναντήσεις Α΄ Η Μεσογειακή διάσταση ενός αιγαιακού νησιού. Β΄ Πέτρα, Πηλός, Ξύλο. Παραδοσιακά υλικά και επαγγέλματα στο Αιγαίο* [Summer meetings in Samos I: The Mediterranean Dimension of an Aegean Island, II: Stone, Clay, Wood. Traditional Materials and Professions in the Aegean] (Athens: Mouseio Physikis Istorias Aigaiou-Palaiontologiko Mouseio Samou-Idryma Konstantinou & Marias Zimali, 2004). See also the seminar organized in 2003 by the NHRF and the University of Manouba (Tunisia) in the context of the Hermoupolis Seminar Series on the subject of *Cultures et produits traditionnels en mer Egée et en Méditerrannée*. See also Eleftheria Zei, "La Terre et la taxe dans la Méditerrannée orientale latine (XIVe-XVIIIe siècles)," *Fiscal systems in the European economy from the 13th to the 18th centuries. Atti della "Trentanovesima Settimana di Studi" 22–26 aprile 2007*, ed. Simonetta Cavaciocchi (Florence: Firenze University Press, 2008), 241–251.

21. Eleftheria Zei, "Ξυλεία και κάρβουνο στο Αιγαίο, 16ος-18ος αιώνας: ελλείψεις των νησιών και συμπληρωματικότητες του Αρχιπελάγους," [Timber and Coal in the Aegean, sixteenth to nineteenth century: Deficiencies and Supplements in the Archipelago], in *Καλοκαιρινές Σαμιακές Συναντήσεις*, ed. Sfoini; eadem, "The Proposal of the Traveller Ch. Sonnini on the Organization of French Commerce in the Greek Archipelago (Voyage in Greece and Turkey, Paris 1801): A First Theory of Economic Insularity," in *Following the Nereids: Sea Routes and Maritime Business, 16th-20th centuries*, eds. Maria-Christina Chatziioannou and Gelina Harlaftis (Athens: Kerkyra, 2006), 61–71.

22. Maroula Sinarellis, "La Mer Egée au XVIIe siècle. Parcours, liens familiaux et recomposition sociale," *Annales. Histoire, Sciences Sociales* 62, no. 4 (2007): 885–918.

23. Constantopoulos, et al., eds., *The Dispersed Urbanity*.

24. Louis Sicking, "The Dichotomy of Insularity: Islands between Isolation and Connectivity in Medieval and Early Modern Europe, and Beyond," *International Journal of Maritime History* 26, no. 3 (2014): 494–511. See also the earlier works of Michel Fontenay on the Mediterranean Christian and Muslim corsair networks, assembled more recently in *La Méditerranée entre le Croix et le Croissant: Navigation, commerce, course et piraterie (XVIe-XIXe siècle)* (Paris: Classiques Garnier, 2010). On the subject of privateer Aegean networks see also below.

25. See for example Nikos Svoronos, "Μια αναδρομή στην ιστορία του Αιγαιακού χώρου," in *Το Αιγαίο Επίκεντρο Ελληνικού Πολιτισμού* [The Aegean as Center of Greek Civilization] (Athens: Melissa, 1992), 33–80.

26. In its initial phase, this concept left out contemporary or previous Latin dominations in the Aegean; later it will be integrated and further elaborated by Spyros Asdrachas and his collaborators in *Ελληνική Οικονομική Ιστορία, ΙΕ΄-ΙΘ΄ αιώνας* [Greek Economic History, fifteenth to nineteenth century], vols. 1–2 (Athens: Politistiko Idryma Omiloy Peiraios, 2005). The book appeared in English translation as Spyros I. Asdrachas et. al, *Greek Economic History, 15th–19th Centuries* (Athens: Piraeus Group Bank Cultural Foundation, 2008).

27. Vasilis Kremmydas, *Ελληνική ναυτιλία, 1776–1835: Όψεις της μεσογειακής ναυσιπλοΐας* [Greek Commercial Shipping, 1776–1835. Aspects of Mediterranean Maritime Navigation] vols. 1–2 (Athens: Istoriko Archeio – Emporiki Trapeza tis Elladas, 1985–6); idem, *Εμπορικές πρακτικές στο τέλος της Τουρκοκρατίας: Μυκονιάτες έμποροι και πλοιοκτήτες* [Commercial Practices in the End of Ottoman Rule: Merchants and Shipowners of Mykonos] (Athens: Nautiko Mouseio Aigaiou, 1993); on eighteenth-century Greek maritime history, see Giorgos Leontaritis, *Elliniki emporiki nautilia 1453–1850* [Greek Commercial Shipping, 1453–1859] (Athens: Etairia Meletis Neou Ellinismou, 1981).

28. Asdrachas, "Patmos."
29. Spyros I. Asdrachas, "Νησιωτικές κοινότητες: οι φορολογικές λειτουργίες (I)" and "Νησιωτικές κοινότητες: οι φορολογικές λειτουργίες (II)" [Insular Communities: their Fiscal Functions], Ιστορικά/Historica 8–9 (1988): 229–258.
30. Eftychia D. Liata, *Η Σέριφος κατά την Τουρκοκρατία* [Serifos during Ottoman Rule] (Athens: Idryma Ereunas kai Paideias tis Emporikis Trapezas tis Elladas, 1987).
31. Sevasti Lazari, "Économies et sociétés des îles de la mer Egée pendant l'occupation ottomane. Le cas de Mykonos" (Ph.D. thesis, Université Paris I, 1990).
32. Evangelia Balta and Maria Spiliotopoulou, "Έγγεια φορολογία και φορολογική απαίτηση στη Σαντορίνη τον 17ο αιώνα" [Land Taxation and Fiscal Obligations in Seventeenth-Century Santorini], *Μνήμων* 18 (1996): 109–142.
33. Dimitris Dimitropoulos, *Η Μύκονος τον 17° αιώνα. Γαιοκτητικές σχέσεις και οικονομικές συναλλαγές* [Mykonos in the Seventeenth Century. Land relations and economic transactions] (Athens: Institouto Neoellinikon Ereunon/ Ethniko Idryma Ereunon, 1997).
34. Maria Spiliotopoulou, "Η Σαντορίνη στην τουρκοκρατία. Κοινωνικές και οικονομικές πρακτικές στο πλαίσιο της οικογένειας," [Santorini during Ottoman Rule. Social and Economic Practices in the Context of family," Ph.D. thesis, University of Crete, 2005.
35. Maria Efthymiou-Chatzilacou, "Rhodes et sa région élargie au XVIIIe siècle: Les activités portuaires" (PhD Thesis, Université Panthéon-Sorbonne, Paris), 1984; Vasilis Kardasis, *Σύρος: Σταυροδρόμι της Ανατολικής Μεσογείου, 1832–1857* [Syros: an Eastern Mediterranean Crossroad] (Athens: Morfotiko Idryma Ethnikis Trapezis, 1987); see also H. Pigné, "Le commerce interinsulaire de l'île de Chios au XVIIIe siècle," *Ιστορικά/ Historica* 8 (1988): 115–122.
36. On this subject, see Socrates Petmezas' critical approach, "La 'commune grecque': une tentative d' histoire des fictions historiographiques," *Byzantina et Moderna: Mélanges en l' honneur d' Hélène Antoniadis-Bibicou*, eds. Gilles Grivaud and Socrate Petmezas (Athens: Melissa, 2007), 207–232. See also the bibliographical survey in Eftychia D. Liata, "Οι ελληνικές κοινότητες (17ος-19ος αι.): από την ιστορία των θεσμών στην ιστορία των τοπικών κοινωνιών και οικονομιών" [Greek communities (18th to 19th century): from the History of Institutions to the History of Local Societies and Economies], in *Ιστοριογραφία της Νεότερης και Σύγχρονης Ελλάδας, 1833–2002* [The Historiography of Modern and Contemporary Greece, 1833–2002] (Athens: Institouto Neoellinikon Ereunon/Ethniko Idryma Ereunon, 2006), 533–540.

37. B.J. Slot, *Archipelagus Turbatus. Les Cyclades entre colonization latine et occupation ottomane*, c. 1500–1718, vols. 1–2 (Leiden: Nederlands Instituut voor het Nabije Oosten, 1982).

38. Aglaia Kasdagli, *Land and Marriage Settlements in the Aegean: A Case-Study of Seventeenth Century Naxos* (Venice: Hellenic Institute of Byzantine and Post-Byzantine Studies, 1999).

39. Nikolitsa Matha-Dematha, "Habitat et rapports socio-économiques a Sifnos (deuxième moitié du XIXe siècle-début du XXe siècle)" (Ph.D. thesis, Université Paris I, 1992).

40. Maria Gianissopoulou, "Société et religion en Grèce insulaire. Un exemple: Potamia-Tinos" (Ph.D. thesis, École des hautes études en sciences sociales – EHESS, 1992).

41. Dimitris Dimitropoulos, *Μαρτυρίες για τον πληθυσμό των νησιών του Αιγαίου, 15ος–αρχές 19ου αιώνα* [Evidence on the Population of the Aegean Islands] (Athens: Institouto Neoellinikon Ereunon/Ethniko Idryma Ereunon, 2003); Nikos Margaris and Lina Mendoni eds., *Κυκλάδες: Ιστορία του Τοπίου και Τοπικές Ιστορίες* [Cyclades: History of Landscape and Local Histories. From the Natural Environment to the Historical Landscape] (Athens: Ypourgeio Perivallontos Chorotaxias kai Dimosion Ergon – Ethniko Idryma Ereunon, 1998); Vasilis Panayopotopulos, Leonidas Kallivretakis, Dimitris Dimitropoulos, Michalis Kokolakis, Evdokia Olympitou, *Πληθυσμοί και οικισμοί του ελληνικού χώρου, Ιστορικά μελετήματα* [Population and Settlements in the Greek Space. Historical Studies] (Athens: Institouto Neoellinikon Ereunon/Ethniko Idryma Ereunon, 2003).

42. Giorgos Tolias, ed., *Το Αιγαίο πέλαγος. Χαρτογραφία και ιστορία, 15ος–17ος αιώνας* [The Aegean Sea. Cartography and History, fifteenth to seventeenth century] (Athens: Morphotiko Idryma Ethikis Trapezis, 2010); Giorgos Tolias and Dimitris Loupis eds., *Eastern Mediterranean Cartographies, 18th International Congress on History of Cartography (Athens 11–16 July 1999)* (Athens: Institouto Neoellinikon Ereunon/Ethniko Idryma Ereunon, 2004).

43. For instance, see Kostas Komis, *Νησιωτικά. Πληθυσμιακές μετακινήσεις, δημογραφικές ανακατατάξεις & οικονομικές διαμορφώσεις* [Of Islands: Population Movements, Demographic Change and Economic Transformation] (Ioannina: Panepistimio Ioanninon, 2004); idem, *Χολέρα και λοιμοκαθαρτήρια (19ος–20ός αιώνας). Το παράδειγμα της Σαμιοπούλας* [Cholera and Quarantine (nineteenth to twentieth century): The Case of Samiopoula] (Ioannina: Panepistimio Ioanninon, 2005); idem, *Σύναμμα. Κοινωνικές Δομές και Όψεις του Νεοελληνικού χώρου: Πόλεις, πληθυσμιακές μετακινήσεις, Μηχανισμοί κυριαρχίας και άλλα σχετικά ζητήματα* [*Summa.* Social Structures and Aspects of Modern Greek Space:

Cities, Population Movement, Mechanisms of Authority, and other Related Issues] (Ioannina: Panepistimio Ioanninon, 2007).

44. Eftychia Kosmatou, "La Population des Iles Ioniennes 18e–19e siècle" (Ph.D. thesis, Paris I-Pantheon-Sorbonne, 2000).

45. Dimitris Dimitropoulos, "Ελαιοτριβεία, μύλοι, φούρνοι, εκκλησίες στο νησιωτικό χώρο τον 17ο αιώνα. Το ζήτημα της συνιδιοκτησίας με βάση το παράδειγμα της Μυκόνου," [Olive Presses, Mills, Furnaces, Churches in Seventeenth-Century Insular Space] *Μνήμων* 16 (1994): 37–70; idem, "Στοιχεία για τον οικιακό εξοπλισμό στα χρόνια της οθωμανικής κυριαρχίας. Η περίπτωση του καθρέφτη," [Information on Domestic Furniture during Ottoman Rule. The Case of Mirrors], *Ιστορικά/ Historica* 24–25 (1996): 37–66; Eleftheria Zei, "Πατητήρι και πιεστήριο: διαφορετικές κοινωνικές ιστορίες του αμπελιού στα νεότερα χρόνια," [Wine and Olive Presses: Vines and their Different Social Histories in Modern Times], *Οίνον ιστορώ. Θλιπτήρια και πιεστήρια. Από τους ληνούς στα προβιομηχανικά τσιπουρομάγκανα* [On Wine IV: Crushers and Mills. From Ancient to Pre-industrial Presses], ed. Giannis A. Pikoulas (Athens: s.i., 2005), 199–213. See also Katerina Korre and Evdokia Olympitou, eds., *Άνθρωποι και παραδοσιακά επαγγέλματα στο Αιγαίο* [People and Traditional Professions in the Aegean] (Athens: Idryma Meizonos Ellinismou, 2003); Dimitris Dimitropoulos and Evdokia Olympitou, *Ψαρεύοντας στις ελληνικές θάλασσες. Από τις μαρτυρίες του παρελθόντος στη σύγχρονη πραγματικότητα* [Fishing in Greek Seas. From the Testimonies of the Past to Contemporary Reality] (Athens: Institouto Neoellinikon Ereunon/Ethniko Idryma Ereunon, 2010). See also Evdokia Olympitou, "La pesca nell' Egeo: La pesca el il pesce nel XIX secolo," in R. Gertwagen, S. Raicevich et al, *Il mare. Com'era. Atti del II Workshop Internazionale HMAP del Mediterranneio e Mar Nero* (Cioggia: ISPRA - Istituto Superiore per la Protezione e la Ricerca Ambientale, 2008), 65–76.

46. Kostas A. Damianidis, *Ελληνική παραδοσιακή ναυπηγική* [Greek Traditional Shibuilding] (Athens: Politistiko Idryma Omilou Peiraios, 1987); idem., "Ναυπηγεία και ναυπηγική τέχνη," [Shipyards and the Art of Shipbuilding], in *Η Ελλάδα της θάλασσας* [Greece of the Sea], eds. Spyros I. Asdrachas, Anastasios Tzamtzis, and Tzelina Harlafti (Athens: Melissa, 2004).

47. See the exemplary work on this domain of Evdokia Olympitou, *Σπογγαλιευτική δραστηριότητα και κοινωνική συγκρότηση στο νησί της Καλύμνου, 19ος–20ός αι.* [Sponge Fishing and Social Development on the Island of Kalymnos, 19th to 20th century] (Athens: Institouto Neoellinikon Ereunon/Ethniko Idryma Ereunon, 2014).

48. Kardasis, *Syros*.

49. Christina Agriantoni and Dimitris Dimitropoulos, eds., *Σύρος και Ερμούπολη. Συμβολές στην ιστορία του νησιού, 15ος–20ός αιώνας* [Syros

and Hermoupolis. Contributions to the History of the Island, fifteenth to sixteenth century] (Athens: Institouto Neoellinikon Ereunon/Ethniko Idryma Ereunon, 2008).

50. Christos Loukos, *Πεθαίνοντας στη Σύρο τον 19ο αιώνα: οι μαρτυρίες των διαθηκών* [Dying in Syros during the nineteenth century: The Evidence of Testaments] (Herakleion: Panepistimiakes Ekdoseis Kritis, 2000).

51. Gelina Harlaftis, *A History of Greek-Owned Shipping: The Making of an International Tramp Fleet, 1830 to the Present Day* (London: Routledge, 1996).

52. Maria-Christina Chatziioannou and Gelina Harlaftis, eds., *Following the Nereids. Sea Routes and Maritime Business 16th–20th Century* (Athens: Kerkyra, 2006)

53. See a global survey in the recent volume Tzelina Harlaftis and Katerina Papakonstantinou, eds., *Η Ναυτιλία των Ελλήνων, 1770-1821* [Greek Shipping, 1770-1821] (Athens: Kedros, 2013).

54. Nikos Belavilas, *Λιμάνια και οικισμοί στο αρχιπέλαγος της πειρατείας : 15ος-19ος αιώνας* [Ports and Settlements in the Archipelago of Piracy: fifteenth to nineteenth century] (Athens: Odysseas, 1997).

55. See also Dimitris Dimitropoulos, "Η πειρατεία στο Αιγαίο. Όψεις και αντιφάσεις των στερεοτύπων" [Piracy in the Aegean. Aspects and Contradictions of Stereotypes], in *Μύθοι και ιδεολογήματα στη σύγχρονη Ελλάδα* [Myths and Ideological Constructions in Contemporary Greece] (Athens: Sholi Moraiti, 2007).

56. Spyros I. Asdrachas, *Ιστορικά απεικάσματα* [Historical Representations] (Athens: Themelio, 1995), 145–148.

57. Asterios Argyriou, ed., *Η Ελλάδα των νησιών από τη Φραγκοκρατία ως σήμερα* [Insular Greece from Latin Domination to Contemporary Times], vols. 1–2 (Athens: Ellinika Grammata, 2004).

58. Asdrachas, Tzamtzis, and Harlafti, eds., *Η Ελλάδα της θάλασσας*.

59. Spyros I. Asdrachas, *Υπομνήσεις. Ιστορικότροπα σημειώματα* [Notes towards History] (Athens: Themelio, 2014). Some isolated pioneer attempts at a history of Greek "micro-insularity" have been made around the middle of the 1990s, as an immediate response to the political attention drawn to the north-eastern maritime frontier between Greece and Turkey and the national debate which had exploded at the time: Leonidas Kallivretakis, *Υπόμνημα περί των νησίδων "Λιμνιά-Ίμια" / Report on the "Limnia-Imia" islets* (Athens: Institouto Neoellinikon Ereunon/Ethniko Idryma Ereunon, 1996); idem., "Ίμια: η αμηχανία περί το όνομα," [Imia: the Embarrassment about the Name], *Ο Πολίτης* 21 (1996), 14–15.

60. Evdokia Olympitou, "Η οργάνωση του χώρου στο νησί της Πάτμου (16ος– 19ος αιώνας)," [The Organization of Space on the Island of Patmos (six-

teenth to nineteenth century)] (Ph. D. Thesis, University of Athens, 2002). See for example the series entitled *Ελληνική Παραδοσιακή Αρχιτεκτονική* [Greek Traditional Architecture] published by Melissa Publications during the 1980s.

61. Christina Agriantoni and Maria Synarellis eds., "Special section," *The Historical Review* 5 (2008), 151–253; also the recent approach of the island of Leukada as a "frontier" of the Ottoman Empire by Elias Kolovos, "Εισαγωγή," in *Οθωμανικές πηγές για τη νεότερη ιστορία της Λευκάδας* [Ottoman Sources for the Modern History of Lefkada], ed. idem, trans. idem and Marinos Sariyannis, trans. (Herakleion: Panepistimiakes Ekdoseis Kritis, 2013).

62. Bernard Vernier, *Η κοινωνική γένεση των αισθημάτων. Πρωτότοκοι και υστερότοκοι στην Κάρπαθο* [The Social Creation of Feelings: Firstborns and Subsequent Borns in Karpathos], trans. Evgenia Tselenti (Athens: Alexandria, 2001). See also the introduction to this book by Eva Kalpourtzi, "Οι σχέσεις συγγένειας ως σχέσεις υλικής, κοινωνικής και συμβολικής κυριαρχίας" [Kinship as Relations of Material, Social and Symbolic Power], in ibid., xi–xix.

63. Peter Loizos and Efthymios Papataxiarchis, eds., *Contested Identities: Gender and Kinship in Modern Greece* (Princeton, NJ: Princeton University Press, 1991).

64. Thomas Gallant, *Experiencing Dominion. Culture, Identity and Power in the British Mediterranean* (Notre Dame, Indiana: University of Notre Dame Press, 2002); see also the Greek translation by Alexandria (Athens 2014); idem, "When 'men of honour' met 'men of law': ritualized violence, the unwritten law and modern criminal justice," in *Problems of crime and Violence in Europe, 1780–2000. Essays in Criminal Justice*, eds., Ephe Avdela, Shani Cruze, and Judith Rowbotham (Lewiston, NY: Edwin Mellen Press, 2010).

65. Evdokia Olympitou, *Μπουμπουλίνα, Καίρη, Μαυρογένους. Οι γυναίκες του Αγώνα* [Women in the Greek Revolution of 1821] (Athens: Ta Nea: 2010); eadem, "Μπουμπουλίνα: Η θηλυκή εκδοχή της ανδρείας," [Bouboulina: the Female Version of Bravery], in *"Η ματιά των άλλων." Προσλήψεις προσώπων που σφράγισαν τρεις αιώνες (18ος-20ός)* [The Gaze of Others. Perceptions of People who Marked Three Centuries (eighteenth to twentieth century] (Athens: Institouto Neoellinikon Ereunon/Ethniko Idryma Ereunon, 2012); Eleftheria Zei, "Les femmes de l' Archipel, XVIIe-XVIIe siècles. Domination féminine: modèles et discours," *Études Rousillonaises* 25 (2013): 163-174; eadem, "Relationships of Affection, Relationships of Power: Death and Family Grieving in the Islands of the Aegean, seventeenth–eighteenth century," *Historein*, 8 (2008): 94–101. On relations of power see also the example of Crete in Ourania Astrinaki, "Powerful

Subjects in the Margins of the State," in *Problems of Crime and Violence*, eds. Avdela et al., 207–236.

66. Eleftheria Zei, *Κανάρης, Κουντουριώτης, Τομπάζης. Ο Αγώνας στη θάλασσα* [Kanaris, Kountouriotis, Tompazis: The Greek Revolution at Sea] (Athens: Ta Nea: 2010).

67. See a diversified recent approach to insular and local elites in Eleftheria Zei and Alexandra Sfoini, eds., *Αυθεντίες. Κύρος και εξουσία στο χώρο της Οθωμανικής και της Βενετικής κυριαρχίας 15ος-19ος αι.* [Authorities: Power and Prestige in Ottoman and Venetian Space, fifteenth to nineth century], Special Issue of *Historica* 59 (2013): 278–471; see especially the analysis of concepts by Spyros I. Asdrachas, "Η Έννοια της Αυθεντίας," 278–281. See also Eleftheria Zei, "Entre la Crète ottomane, l'Égypte et la France: le cas de Konstantin Kozyris à Candie sous l' administration égyptienne (première moitié du XIXe siècle," in *Hommes de l'entre-deux. Parcours individuels et portraits de groupes sur la frontière de la Méditerranée (XVIe-XXe siècle)*, eds. Bernard Heyberger and Chantal Verdeil (Paris: Les Indes savantes – Rivages des Xantons, 2009), 219–229; eadem, "Relationships of Affection."

68. Nicolas Vatin and Gilles Veinstein, eds., *Insularités ottomanes* (Paris: Maisonneuve & Larose, Institut francais d' études anatoliennes, 2004).

69. Eleftheria Zei, *Paros dans l'Archipel Grec, XVIIe-XVIIIe siecles: les multiples visages de l' insularité*, Istanbul: Isis, forthcoming); eadem, "Πρώτες προσεγγίσεις στη διαμόρφωση τοπικών 'ελίτ' στο Αιγαίο του 18ου αιώνα : οι 'άρχοντες' του μακτού," [A First Approach to the Development of Local Elites in eighteenth-century Aegean: the Notables of Maktu], in *Αυθεντίες*, eds. eadem and Sfoini,385–398.

70. Elias Kolovos, *Η νησιωτική κοινωνία της Άνδρου στο Οθωμανικό πλαίσιο. Πρώτη προσέγγιση με βάση τα οθωμανικά έγγραφα της Καϊρείου Βιβλιοθήκης (1579–1821)* [The Insular Society of Andros in the Ottoman Context. A First Approach Based on the Ottoman Documents of the Kairios Library (1579–1821)] (Andros: Kaireios Vivliothiki, 2006).

71. See the following essays on the Duchy of Naxos: G. Saint-Guillain, "Cavalieri, feudatarii, borghesi e altri vassal": Le forme di notabilità nelle signorie delle Cicladi (13o-15o secolo); A. Kasdagli, "Memoria collective e realtà sociale. Le famiglie nobili di Nasso dopo la caduta del ducato"; B. J. Slot, "Il Mito del ducato dell' Arcipelago," in *Το Δουκάτο του Αιγαίου* [The Duchy of Aegean], eds. Nikos G. Moschonas and M.G. Lily Stylianoudi (Athens: Instituto Neoellinikon Ereunon/Ethniko Idryma Ereunon, 2009), 136–148, 447–454, and 466–475 respectively.

72. See for example the first publication of the results of the academic research program on Thirassia, the islet opposite Santorini, realized with the collaboration of the University of Crete (Faculty of History and Archeology), the University of Salonica (Faculty of Architecture) and the Ephorate of

Antiquities of the Cyclades: Klairi Palyvou and Iris Tzahili, eds., *Θηρασία I: μια διαχονική διαδρομή* [Therassia: A Timeless Trajectory] (Athens: Ta Pragmata, 2015); the archaeological field work on the islet of Gavdos (Crete) and the interdisciplinary research team of the University of Crete under the supervision of Katerina Kopaka, Professor of Prehistoric Archaeology, which has organized a first meeting on the islands in December 2014 in Rethymno (proceedings to be published); see also the papers presented at the above-mentioned *Insularities Connected* international conference recently organized in Crete.

PART II

Violence and Law in Terraqueous Space

Separating the Waters from the Sea: The Place of Islands in Ottoman Maritime Territoriality during the Eighteenth Century

MICHAEL TALBOT

This article is an attempt to situate islands as a part of Ottoman claims of maritime territoriality, focusing on the Eastern Mediterranean in the eighteenth century. That is, I hope to show that the sorts of ways that officials of the Ottoman state wrote about issues of Ottoman control in certain bodies of water demonstrate that pieces of land denoted as islands—*aḍalar* or *cezāyir*—performed a particular role in territorial delineation. This role, however, was not one of a specific maritime border. Rather, islands acted as one distinct element of a coherent littoral frontier separating the Ottoman Empire from the open sea and its many dangers, and as such were thought of in the same way as ports and coastal fortifications. As with ports and forts, islands were centers of the projection of territoriality and legal authority over surrounding waters, even if they themselves were only loosely subject to the sultan's authority. They created a space both on land and in the surrounding waters over which the Ottoman state claimed practical legal control. More than this, islands were one element of a wider littoral landscape that defined what the Ottoman Empire considered to be its waters in the Mediterranean, making them one incarnation of a particular set of spaces through which interactions with outsiders were regulated and imperial authority asserted. Territoriality was therefore a central part of Ottoman insularity, where islands joined coastlines in forming the limits of the Ottoman state.

Using examples of imperial commands from a wider body of over two hundred archival documents concerning Ottoman attempts to maintain peace and order in what they considered as their waters, with regard to both foreigners and their own subjects, I will argue that the idea of coastal protection was central to the Ottoman understanding of the space around their islands. Territoriality was not an abstract claim, but signified the ability of the Ottoman state to assert its law and justice over a particular maritime space that existed between the shore and the open sea.[1] With their extensive coastlines, islands provided a particular sort of space within which this ability was tested, in particular by the actions of local and foreign pirates attacking mercantile shipping in the surrounding sea. The usual Ottoman response to these threats was to seek to protect the islands as a whole, land and sea, as the security of the maritime traffic that passed islands was equally important to the imperial duty of protection as was that to the islands' inhabitants.

Foreign Challenges and Extreme Limits

The Ottoman state faced numerous security threats in the Mediterranean from its own subjects, and combating such challenges involved dispatching ships or garrisons to the affected areas to secure the lives and property of local inhabitants. This is certainly an important aspect of Ottoman maritime territoriality, and of Ottoman imperial authority. In 1183 (1769), the governor of Rhodes made a petition to the Ottoman government requesting a force of galleons to destroy the pirates, and the appointment of a military official to ensure safe travel in the Mediterranean.[2] Consequently, it was commanded that, in addition to the imperial galleons already patrolling the Ottoman Mediterranean, the galleon *Tuḥfetü'l-mülūk* be sent with a full compliment of *levend*s (naval military forces), and that a *başbuğ* (a military commander) be dispatched to oversee the operation.[3] Such expeditions seem to have been a fairly regular occurrence throughout the eighteenth century, despite the effort and cost involved.[4] In essence, it was as great an imperative that those vessels and officials be dispatched to restore order in the seas around Rhodes as it was for imperial soldiers to be sent to quell rebellion in distant land centers like

Baghdad; this was, after all the basis for imperial authority.[5] As Nicholas Vatin noted in his analysis of imperial control in sixteenth-century Basra, frontier territories could be ruled in different ways, but territories "conquered but not vassal in no way merit the title 'well-protected.'"[6] The one word that occurs again and again in documents concerning piracy and maritime matters is *muḥāfaẓa*; protection or defense. It was this ability to protect, regulate, and control that marked the coastal *ṣu* (water) from the open *baḥr* (sea), and, as firm features on the maritime landscape, islands were crucial to defining and implementing this distinction.

Ensuring that the Well-Protected Domains remained well-protected meant not just tackling local pirates and brigands, but dealing with troublesome foreigners. Of course, this meant utilizing coastal fortifications, the imperial fleet, and corsairs to protect Ottoman seas and ports from assault by outright enemies, be they Venetian, Habsburg, or Russian. Yet, the waters of the Eastern Mediterranean also played host to the warships and privateers of Ottoman friends engaged in conflict throughout the eighteenth century. These visitors—generally British and French—were neither welcome, nor well-behaved, fighting under Ottoman fortresses and blockading harbors, not to mention causing great damage to the Ottoman economy by assaulting the ships used to freight goods between Ottoman ports, hitting staples such as wheat, coffee, and soap particularly hard. The attacks on shipping in particular posed a major challenge to Ottoman authority in the Eastern Mediterranean, and required firm action from the state. Of course, military action was not an option, as the offending parties were states holding *'ahdnāme*s (Capitulations) from the sultan and whose subjects were *müste'min*, protected foreigners. Thus, from the first major attacks in the Aegean in the late 1690s, which saw naval battles around Izmir and prizes taken within island harbors, the Ottoman state instituted what it later called the *şurūt-u deryā*, the sea or maritime regulations.[7] In 1696 in order to stop these assaults, an imperial decree was issued forbidding armed ships of European powers to pass a line going from Andros to Kos and ending at Foça (near Izmir), and ruling that offending parties would be obliged to pay compensation for any prizes taken beyond that maritime border. This was reinstituted in 1703, going from the Anatolian coast to Samos, Ikaria, Andros, and ending at Euboea. During the greater violence of the mid-eighteenth century, these maritime regulations were extended and reinforced in 1744, 1758, and 1779 on an enlarged scale. In 1744 and 1758, the limits of the Ottoman maritime territory were given as from the south of Morea to the Gulf of Sirte in Libya, and in 1779 as being a line between the Morea, Crete, and the western border of Egypt. There is no explanation as to why Crete was included in the 1779 regulations, and not the earlier ones. Perhaps European accusations that the maritime border was "chimerical" necessitated a more fixed position, thus giving Crete as an island a very important role at this stage in defining Ottoman maritime territoriality. Regardless, these regulations, which are mapped in Figure 1 below, saw the Ottoman state extend its maritime borders out of the Aegean, cutting off the entire Eastern Mediterranean from foreign privateering. What is more, the Ottomans successfully enforced these new boundaries, securing compensation for Ottoman merchants who lost their goods on board ships attacked by the French and British privateers who broke the regulations.

Little has been written about how the Ottomans thought about the sea in a legal sense, yet from the orders and regulations drawn up to curb attacks by foreign and technically friendly privateers we can get a real sense as to the thought process behind Ottoman policy. One example of such an imperial command issued to governors and naval officials on 20 Safer 1193 (9 March 1779) was aimed at nipping the situation in the bud. With Britain and France once again at blows due to the war in North America, the privateers of those two nations returned to the Mediterranean in force. Following complaints by French merchant captains of attacks on their ships, the command was issued to bring back the regulations with a slightly modified geographical route, but with the same principle:

> Enmity and contention having arisen among the European states, and in consequence of the faithful observance of the former sea regulations being neglected in various ways by the ships of the said states, a line is to be imagined going from the realm of Morea to the island of Crete.[8]

This *ḥaṭṭ-ı mefrūż*, the "imagined line," marked the outer limits of Ottoman maritime territoriality, within which European ships could not attack each other or others. It is no coincidence that islands helped to form

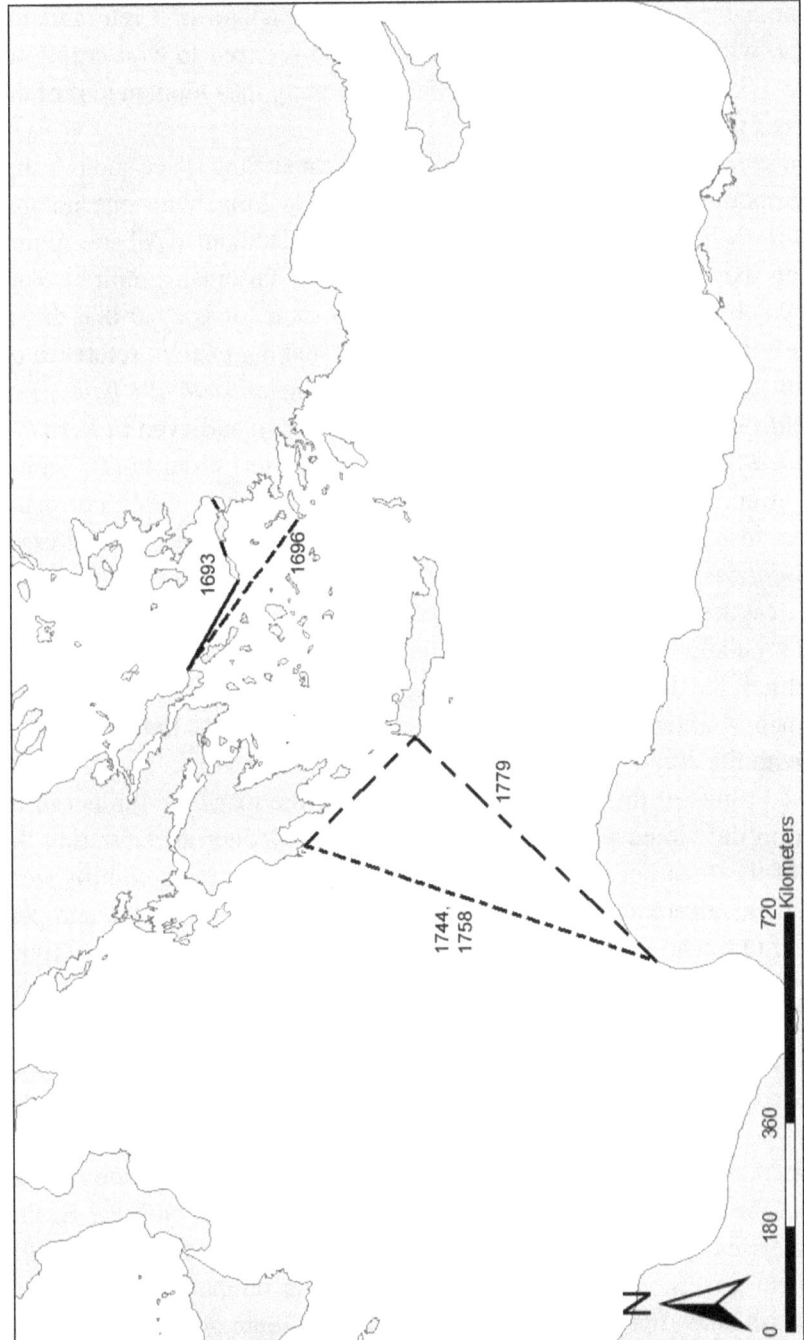

The Ottoman maritime border in the Mediterranean as defined by the 1696, 1703, 1744, 1758, and 1779 regulations

this boundary in the north, between the large island of Crete and the Morea, which could also, as will be seen, be referred to as a *cezīre*, an island. Islands clearly played some part in forming the Ottoman idea of the limits of its liquid territory.

These regulations introduced notions about maritime space and law that contemporaries assumed the Ottomans to be lacking, with one notable maritime scholar of the early nineteenth century, Domenico Alberto Azuni, writing that he knew of no maritime law in the Ottoman Empire except that which had been introduced by the Europeans.[9] It is clear that this is patently false, with the 1193/1779 regulation making explicit reference on several occasions to an Ottoman understanding of *kavāʿid-i bahriye* or *kavāʿid-i deryā* (maritime rules or rules of the sea), and even of *kavāʿid-i merʿīye-i deryā* (rules of respect at sea). They set territorial limits, stating that, in normal circumstances, British and French ships could not make attacks inside the ports or under or in front of the cannon of fortresses (*limānları derūnlarında ve kılāʿ topu altlarında ve pīşgāhında*), nor within three leagues of the coast (*sevāhilinden*). Thus, in the case of the 1193/1779 regulation, the repeated violations of Ottoman ports, fortresses, and coastlines led to the introduction of a blanket ban on violence in the Ottoman Mediterranean, extending the limits of Ottoman authority out to sea, with the aim of protecting land and maritime territory.

The origins of this particular incarnation of the maritime limits can be found in the violence that plagued the Eastern Mediterranean during the War of the Austrian Succession (1740–48). Here, the main culprits were British privateers attacking French shipping carrying the goods and persons of Ottoman subjects. Such traffic was crucial to the internal activity of Ottoman merchants freighting goods around the Mediterranean, and above all in the lucrative routes between Egypt and the North African Regencies, and the Levant and Aegean. Mapping out the attacks that occurred between 1744 and 1747 reveals to what extent the islands under Ottoman control were areas around which this privateering violence occurred, partly because they formed important landmarks for commercial traffic. From the petitions made by Ottoman subjects to both the British embassy and the Ottoman authorities, and the commands issued by the Ottoman government, it is possible to track the rampages of a number of British ships that took French ships with Ottoman goods, with two in particular—*The Ruby* and *The Fame*—being the worst perpetrators. In 1744, *The Ruby* made at least four captures involving Ottoman subjects, off Kythera, Crete, Rhodes, and Zakynthos. In 1746 and 1747, *The Fame* made attacks off Crete, and along the Syrian coast and Cyprus. Both ships, in targeting French ships, managed to take cloth, soap, and—most damningly from the Ottoman perspective—much-needed shipments of rice and coffee being sent to Cyprus and Syria from Egypt. In both cases, the attacks centered on islands, in the west around Crete and the entrance to the Aegean, and in the east around Cyprus. In issuing maritime regulations that made the Mediterranean east of the imaginary line between Morea and Sirte not just a *mare clausum* but, in essence, an extension of the land in which Ottoman laws would be in full force, the Ottoman state was able to exert enough legal and diplomatic pressure to secure compensation or restitution for their wronged subjects from the British authorities, amounting to over 80,000 *guruş*.[10]

The Ottoman commands issued during this period give us an excellent sense of the state's understanding of its maritime space. It was at this time that the Europeans were first forbidden from sending armed ships beyond an imagined line from Mora to Sirte in Libya. A command issued to the governors and commanders in the Ottoman islands, as well as to Ottoman naval commanders at the end of Muharrem 1157 (mid March 1744) reported complaints of the French that one of their ships had been taken by pirates cruising around the islands of Cephalonia and Aya Mavro (Lefkada), and elsewhere in "the waters of the island of Morea" (*Mora cezīresi şularında*).[11] As Kahraman Şakul illustrates in this volume, the Morean peninsula was *almost* an island, with only a narrow joint at Corinth connecting it to the rest of Greece.[12] By emphasizing the insular nature of the Morea, it was placed on the same level as both large islands like Crete and smaller ones like Lefkada. In other words, the recipients of the order—who comprised various ranks of Ottoman official throughout the Mediterranean islands—were being asked to think of all sorts of ports, islands, and coastal landmasses as a common frontier under threat. Thus, when the command spoke of forbidding European ships from coming within "a number of leagues" (*birkaç mīl ba'īd*) of Ottoman territory in accordance with the "ancient regulations of the sea" (*ḳā'ide-i ḳadīme-i deryā*), the islands and ports (*cezāyir ve limānlar*) that had been attacked and would now be protected, formed a homogenous space together that defined Ottoman territorial waters proper (*şular*).

This creation of a united insular and litoral territory helps us to understand what was perhaps the most important of the surviving orders from this period, issued at the end of Şaban 1157 (beginning of October 1744) to Ottoman governors and judges around the Mediterranean coast and islands.[13] The opening narrative described the attacks by friendly foreign states against Ottoman subjects and their goods. The solution was, as in the earlier conflicts, to draw a line. But here, the language is quite different to the later decree, which, as we have seen used the term *ḫaṭṭ-ı mefrūż*, the imagined line, to describe the new maritime border. Here, something more fixed is described that would deter any further raids (*akın*) against Ottoman waters (*Devlet-i ʿAliyem şularında*), specifically commanding that "a boundary be fixed in the sea" (*deryāda bir ḥadd taʿyin olunub*). The purpose of this boundary (*ḥadd*) was to stop European ships fighting each other "in the waters of my Sublime State and in the open oceans, on the shores of Rumelia and Arabia [i.e. Europe and Asia]" (*Devlet-i ʿAliyem şularında ve açık enginlerinde ve Rūm-ili ve ʿArabistān kıyılarında*). This command makes several mentions of the protected spaces from foreign attack, such as "the waters of my Well-Protected Domains, under the guns of the castles, and around the islands" (*memālik-i maḥrūsem şularında ve ḳalʿe ṭopu altında ve cezāyir civārında*). Once again, we find divisions between different sorts of maritime space, separating the shores and territorial waters from the open sea, with islands forming a distinct part of this maritime frontier. Yet, all of these spaces—coastal waters, harbors, coastal fortresses, the waters around islands—are littoral features, and it is perhaps this kind of space that helped to define territoriality across the maritime frontier.

Borders and Frontiers, Solidity, and Fluidity

Defining the physical limits of early modern empires is often a difficult and problematic task. It has often been the case in recent years for the terms "fluid borders" or "fluid frontiers" to be employed in seeking to understand what Palmira Brummett termed as the "large and porous" frontiers of the Sublime State.[14] Describing a border as "fluid" is to describe it precisely as it should *not* be in geopolitics, that is, changing and changeable. Thus, the images invoked by the concept of fluidity in the borders of the state are perhaps problematic in their assumptions when speaking about certain periods.[15] An n-gram query for "fluid borders" in one major online scholarly repository reveals that usage of that term skyrocketed from precisely 1989. The implication that this was a specific reaction to the fall of the Soviet Union, and the idea that borders had been broken down and new opportunities for interactions had opened up, remains a powerful one in scholarship on that period.[16] If the (re)gaining of fluidity could indicate the situation of post-Soviet Europe, then the loss of fluidity through the fixing of permanent borders following the treaty of Carlowitz and subsequent agreements has certainly been employed to explain changes in the nature of the Ottoman Empire from the beginning of the eighteenth century.[17] If part of the parameters of our historical enquiry is to define the space of our subject, then the solidification of borders must surely be a pivotal moment.

Yet, if we frame our notion of space in terms of states being solid or fluid, how do we deal with borders that actually possess those physical properties? Does a mountain range restrict fluidity compared to a river? Is a woodland particularly porous as a border?[18] And then, what about the sea? Is it even possible to speak of the sea as having any part in defining the borders of a state? In fact, understanding attitudes towards the sea is crucial to understanding notions of territoriality. In his extended discussion of the history of international relations in East Asia, David Kang opened a chapter on nomads and islands with the assertion that "the difference between a border and a frontier is the difference between a line and a space."[19] For Kang, a border was a fixed dividing line between two polities, whereas a frontier represented a state of political fluidity, which could be turned solid through the expansion of fixed authority into the area. The term "frontier," however, has its own problems, not least when thinking about Ottoman history with its connotations of perpetual warfare and lack of imperial control.[20] To briefly reclaim the term for the purposes of this paper may well prove problematic, but it deserves some consideration, based on the work of Hugh Clark on ideas of the frontier in Chinese history, in which he defined the frontier as a space of cultural division.[21]

In his discussion on frontier discourse, Clark provided a most eloquent articulation of the sea as a border:

> For all its mystery, however, the maritime frontier was real, and the reality of this frontier was different from that of land frontiers. Because at first glance they are so precisely demarcated, they are both a frontier and a boundary. They appear to offer no abstraction. When one reaches the shore, that is the boundary, and the boundary is the frontier. There is no immediate "other side." What is there is the empty ocean. Thus the maritime frontier appears to be unlike any land frontier; where the latter are imprecise, the maritime frontier presents a frontier that is very clearly defined.[22]

This, in Clark's argument, gave a maritime frontier a very particular set of properties. If a land frontier was a place of cultural mingling, then the maritime frontier was an "interface," a space where two cultures could meet, but where the interaction could be controlled and regulated. From this, we have an intriguing framework within which to think about the sea as a means of defining territoriality: a clear border at the shore; and a means by which to regulate interactions with others.

Thinking about the Ottoman case, and islands in particular, requires reworking this framework somewhat. The idea of the interface is useful, as the maritime frontier was a place that housed the docks, customs houses, and defense networks that marked that space as Ottoman. However, in the Ottoman understanding of territoriality—as with many others in Europe at the same time—the shore and its institutions did not form the border in itself, but were a marker for the actual limits of territorial authority, the territorial waters that occupied the space between the shore and the sea. That is, the borders of the Ottoman Empire in this case were quite literally fluid. Placing islands within this framework complicates the assumption of the idea of interface in this regard. Islands, of course, are pieces of land entirely surrounded by the sea; this is, as Michel Fontenay reminded us, what makes islands specifically different from other sorts of land.[23]

This definition renders islands as distinctly littoral entities. The shores of the islands cannot be separated from their waters nor their hinterland, defined as much by their liquidity as by their solid state. Histories of coasts and littoral spaces have started to make important contributions to conceptual notions of maritime space in general, and the engagement of scholars such as Isaac Land and David Worthington in developing the field has led to interesting collaborative work.[24] One of the questions that Land raises in his writing is the utility of coastal history as a sub-field, and in a review essay from 2007 he provides a compelling and rather beautiful argument in its favor:

> "Oceanic" history was always a metaphor: how many historians ever wrote about salt water? Coastal history is a more productive, and instructive, metaphor. Coastlines would not exist without their proximity to the ocean, but their character is not determined solely by the ocean's action. Coasts may form bulwarks of resistance to the waves, as in the case of coral reefs or towering cliffs. Yet there are messy, intermediate places like tidal flats and brackish estuaries. There are also quite coves and inlets, connected to the ocean but only gently shaped by it. [...] In their diversity, and in their ever changing nature, coasts parallel the diverse experiences of human beings in their confrontation with water, and each other.[25]

The variety and diversity of the coastline, its permanence and transience, is an important factor in accounting for and interpreting its role in human politics and society. The transience of permanence is an important idea, one which the geographer David Harvey defines as a perpetually perishing space dependent on processes, usually capital flows.[26] The relationship between coastal space and human is therefore contingent, and translates the space beyond a mere border between the blue ocean and the green grass beyond. John Gillis has argued for a complication in our understanding of littoral spaces, borrowing an ecological term to describe the coast as an ecotone, where "land and water constitute and ecological continuum."[27] In thinking about coasts not as forming a border between land and sea, but as a discrete space in which the two are merged into something distinct, then the character of islands becomes rather interesting, particularly in terms of their territoriality.

As points of land surrounded by sea, one way of thinking about islands is as nodes of territoriality, extending the authority of the state out into the open sea. This is achieved, in part, through their ecotonal nature, merging and forging a new space from water and land. In this sense, the Ottoman imperial commands that we often find islands and territorial waters together almost as synonyms, an demonstration perhaps of the deep significance of what Gillis calls "brown water history".[28] In a command to the *Kapudan-ı Derya* in 1758, representative of a number of similar commands issued in the second half of the eighteenth century, the admiral was instructed to patrol "around the islands and other waters of my Sublime State" (*aḏalar aralarında ve sā'ir devlet-i ʿaliyem ṣularında*).[29] The spatial implication here is that islands are a distinct but connected part of Ottoman territorial waters. What, then, of archipelagos or other clusters of islands? For instance, in the Eastern Mediterranean, Cyprus is a lone island between Anatolia and the Levant surrounded by hundreds of miles of open sea. Crete, however, sits at the entrance to the Aegean, which is broken up by dozens of islands, each one acting as a small node of imperial authority in their surrounding waters, assisted by what Cyprian Broodbank has termed "a high degree of inter-visibility."[30] Does this intensification of territoriality through the presence of many islands then indicate a stronger Ottoman claim to territorial authority in the Aegean than it could make in the Levant? In other words, how far did Ottoman maritime territoriality depend on islands separating Ottoman waters from the sea beyond?

Sultan of the Two Seas?

From examining a number of imperial commands concerning piracy and maritime violence in the Mediterranean Sea in the seventeenth and eighteenth century from the perspective of the Ottoman state in Istanbul, it seems that the sea was divided into two parts. The first was *rūy-u deryā*, literally "the face of the sea." This term was used to describe the open sea, the waters considered outside of direct imperial control. To give an example, at the beginning of Rebiülevvel 1122 (beginning of May 1710), a command was issued to İbrahim Pasha, one of the Ottoman galleon commanders in the Mediterranean.[31] It reported that on 11 Muharrem of that year (March 12, 1710) a pirate galleon and two saitees appeared from the open sea (*rūy-u deryā...ẓuhūr*), and began targeting shipping between the Egyptian commercial hub of Damietta and the Syrian ports such as Sidon, Tripoli, and Payas. The Ottoman solution was to send two galleons under the command of İbrahim Pasha to patrol between Payas and the straits of Damietta, and this was really all they could do to pursue their aim of protecting Muslim ships from pirate attacks (*müslümān sefinelerini ḳorsan eşḳiyası mużırrından emin ve sālim eydüb*). The open sea between Egypt and Syria was just that; open, featureless, ungovernable. The language used in this report, that the pirates appeared or emerged (*ẓuhūr*) from out of this vast expanse, perhaps gives a sense of how this space was perceived. These ships were only comprehensible when sighted, and more often than not such an encounter would happen in or around coastal waters rather than in the open sea, as projections of power and control emanated from the coastal defenses and ports.[32]

Mounting these defense operations was an essential duty, as the claim to territory, both rhetorically and practically, was crucial to the legitimation of the Ottoman state. Among the various grandiose titles claimed by the House of Osman, one of the more frequently employed phrases in monumental epigraphy and diplomatic correspondence was *sulṭānü'l-berreyn ve ḥaḳānü'l-baḥreyn*, the Sultan of the two lands and ruler of the two seas. Histories of the Ottoman Empire have focused extensively on the first element of this title, exploring the theories, methods, successes, and problems of the assertion of Istanbul's authority over the urban and rural space of the imperial dominions in Europe and Asia. Indeed, much of our understanding of what made the Ottoman Empire what it was has been shaped by its great landmass. However, great strides have certainly been made in understanding the Ottoman relationship to the sea, upon which the assertions of this present study are entirely dependent.[33] The second part of that grand imperial title is crucial in developing our understanding Ottoman state and society, its place in the world, and its self-image. As well as claiming its two great seas—the Black and the White (the Mediterranean)—rhetorically, at various different times the Ottoman state also attempted to exert various levels of imperial control over them. Beyond this, we need to consider imperial influence in the Red Sea, the

Persian Gulf, and the Indian Ocean, not to mention the innumerable bodies of water inland, the rivers, and lakes. Moreover, with so much of the population of the Ottoman Empire being riverine, lacustrine, or littoral, and with the sea playing such a central role in trade and travel, we simply must consider the Ottoman relationship with liquid space in order to fully appreciate the nature of imperial authority.

One of the reasons, I suspect, that the sea is often passed over in analyzing the Ottomans as an Empire, is because it is difficult to envisage authority over something that is, quite literally, intangible. Unlike rivers and other inland bodies of water, the open sea, as the early modern European theorists of international law posited, cannot be so easily physically occupied.[34] A warship can only control a limited part of a vast and ever-shifting expanse, and then only temporarily, dependent on finite supplies and fickle weather. Maritime powers could use their naval might to project such control over sea-lanes, but these lanes were not physical highways, but ill-defined and temporary passages from one landmass to another, the very essence of Harvey's permanences. Perhaps the only analogous space to that of the abstract sea in terms of imperial authority is the abstract desert, which is also a large and largely featureless space, with human habitation temporary or transitory. Taking this analogy further, we might begin to think about the old cliché of oases being islands of life in the desert, and consider islands to be oases of the sea, without which control would be impossible. Yet, did the Ottoman claim of maritime sovereignty and territoriality rest on authority over the islands in the seas it sought to control? Was Ottoman maritime territoriality, in effect, insular?

This leads me to the parts of the Mediterranean immediately adjacent to Ottoman shores that formed the second element of that sea, generally referred to simply as *şular*, the Ottoman "waters." These were controllable and crucial, the waters extending "*birkaç mīl*," a certain number of miles (usually up to three) from the shore mentioned in maritime regulations. Over half a century after the attacks on the Egypt-Syria shipping, a command was given in the middle of Şeval 1179 (end of March 1766) to the *Kapudan-ı Derya* Hüseyin Pasha concerning the security of trade in the Mediterranean.[35] The order complained of the banditry of pirates (*eşkiya-yı ḳorṣan*) in attacking merchants and protected foreigners (*tüccār ve müste'mīn ṭā'ifesiniñ*), who were going to Egypt and travelling in the "waters of my Well-Protected Domains in the Mediterranean Sea" (*baḥr-ı sefīdiñ memālik-i maḥrūsem ṣularında*). The complaints had been made by the governors and commanders resident in the islands and on the coasts (*cezāyir ve sevāḥilde*), who were evidently concerned at the effect this maritime violence would have on their territories. What was important in this case was that the violence on this occasion was happening in and around coastal and island waters, and not the open sea. The Ottoman solution was to dispatch the imperial fleet that summer to Morea, "around the islands and the other waters of my Sublime State" (*aḍalar aralarında ve sā'ir devlet-i ʿaliyem ṣularında*). Here, in a verbatim repetition of the command of 1758, the islands form a particular element of the maritime landscape, a specific part of the "waters" under Ottoman control, not separate legally or territorially, but important practically as markers for defining liquid imperial authority, an ecotone of territoriality and maritoriality. Yet it was, as the command also tells us, the *cezīreler sāḥilleri*, the coasts of the islands, and the populations living there, that made their protection so important, and we might also see those coasts as providing the islands' ecotonality.

Given the importance of islands in defining these territorial waters, we should not discount the relationship between Ottoman maritime territoriality and insularity per se. To consider one extreme that largely removes islands from the equation, the maritime border imposed by the line drawn between Greece and Libya, intended to block the Eastern Mediterranean in its entirety was a successful assertion of maritime authority when the violence of European allies proved a significant security challenge and a profound threat to Ottoman shipping. This demonstrates, quite conclusively, that in the eighteenth century the Ottomans refused to be passive observers of violence in their claimed waters, and were capable of dealing with these threats (when posed by nominal allies) without recourse to violence themselves. In her pioneering study on the Ottoman sixteenth century, Palmira Brummett demonstrated that imperial space was relative: it could be practical through the collection of taxes or levying of troops; rhetorical, in that there was no guarantee that authority in a particular region would be respected; or imagined as a means to intimidate opponents.[36] Based on this, we could well say that the Ottoman maritime lines comprised a bit of all three categories. They created a maritime space that was largely imagined because of a lack of ability to physically control or divide the sea, employed the maritime regulations as a rhetorical device to dissuade foreign warships from making attacks, but ultimately that was practical because of the successful prosecutions made against foreign privateers. In other words, Ottoman maritime territoriality in the Mediterranean was not necessarily as abstract as the boast "Sultan of the two seas" *sulṭānü'l-baḥreyn* initially suggests.

These limits were, however, also entirely divorced from the island and littoral landscape. As with many pieces of historical evidence, we only find out about Ottoman attitudes towards their maritime space when things went wrong. The maritime limits were a temporary measure. The imagined line was a legal ploy for a specific and challenging context, not a permanent assertion of Ottoman legal authority over the entire sea; once the trouble abated, the regulations were discarded until the next conflict. As such, the maritime limits, whether they were fixed or imagined, lines or borders, were not comparable to the fixing of land borders in the same period. Ultimately, the Ottoman state in the eighteenth century was not interested in asserting a universalist claim over the Mediterranean. The *rūy-u deryā* or *açık engin*—the open sea—remained a non-territorial space. When imperial commands do refer to *açıklar*, open waters, in the context of territorial waters, this refers almost exclusively to the maintenance of shipping lanes. Therefore, to consider maritime territoriality aside from this extreme regulation, we need to return to a more literal landscape.

That the fixing of land borders and the articulation of maritime space occurred in the same period is not a coincidence. The processes, however, were entirely different; this might be a trite statement, but it is necessary to articulate these differences.[37] The Ottoman archives are full of little sketches, maps, and textual descriptions of border commissions and negotiations. To take one example, after the Treaty of Sistova (1791) it was necessary to settle the route of the Ottoman-Habsburg border. One of the few places transferred from the Ottomans to the Habsburgs was a parcel of land in between the Una and Glina rivers (today the northwest border between Croatia and Bosnia-Herzegovina). A standard map was drawn up as per the treaty to show the new frontier; it provides a number of features used to define the border, including rivers, streams, mountains, villages, and castles.[38] In this case, a line could be drawn on a map, agreed by both parties, marked by set features, and henceforth it would be understood that on one side of the line was Habsburg territory, and that on the other was Ottoman. Thus, when foreign merchants crossed into the Ottoman Empire by land, there was, at least on a geopolitical level, a moment when they could say for certain that they had set foot upon Ottoman territory.

The processes of defining the Ottoman land borders occurred largely in the context of treaty negotiations with belligerent neighbors, something that cannot be said for territorial waters in the Mediterranean. However, for merchants coming by sea, there was also such a moment when they could be said to have reached Ottoman territory at sea. On first consideration, it might appear that their experience of being in Ottoman territory, enjoying their freedoms as *müste'min* and becoming subject to the consequent financial and legal regime, began only on their arrival in port. From another perspective, it could be understood that islands formed natural physical boundaries marking the start of Ottoman waters, at least in the minds of these historical actors. Passing Kythera, Crete, or Rhodes would take merchants into a sea bounded by Ottoman territory on all sides and filled with Ottoman islands. Perhaps, then, the Aegean islands could be understood as equivalent to forests or rivers on land in marking Ottoman territory. This, however, does not seem to be the Ottoman understanding of their territoriality as expressed in the documents I have examined: after all, as a number of the commands demonstrated, they considered their authority to stretch just beyond the coast, a number of leagues—*birkaç mīl*—out into the sea. The part of the sea that required protection was the *şu*, the territorial waters, and violations there called for action from the state. There *was*, therefore, a tangible moment when a ship came into the Ottoman Empire's territory on the water and was entitled to receive Ottoman protection, and the coasts of both islands and the main landmasses helped to define this, not least because of the authority they projected through their nature as concurrently liquid and solid space.

Conclusions

Was there something particular about the place of islands in Ottoman territoriality? Gilles Veinstein said that in studying Ottoman insularity, we are attempting "to discern what specific place and role this state reserved to its possessions of a certain kind that distinguishes them from all its other provinces, themselves diverse in nature."[39] Does the Ottoman approach to maritime territoriality in the eighteenth century distinguish islands from other spaces, such as coasts, castles, or ports? I would argue, in this case, that this was a territoriality defined by things littoral rather than particularly insular. Islands were a crucial part of defining this liquid territory, but this did not make the Ottoman claim to the Aegean all encompassing; in terms of territory, it was not an Ottoman lake, nor an engorged river populated by islets. Just a few leagues beyond an island's shores, territorial waters became open sea once again, outside of imperial control and legal authority. However, whilst the maritory of the Aegean islands was defined by their coastal waters and not simply by virtue of their being in the Aegean, their role in defining space within that open sea was crucial.

Most importantly of all, that definition of space had a purpose beyond the bluster of official rhetoric. If we return to one of the cases examined earlier, that of the Ottoman attempt to stop pirate attacks between Syria and Egypt in 1710, we can see the rationale behind claims of maritime territoriality and their articulation. In justifying the deployment of the imperial fleet in force—with galleons, galleys, and frigates—to patrol the open waters between Damietta and Payas as we saw above, the command explained that it was aimed at securing justice for "the poor subjects and people of the realm residing on the coasts of all the isles and islands" (*bilcümle aḍalar ve cezāyir sāḥilinin sākin olan fuḳarā-yı raʿiyyet ve ehl-i memleketiñ*).[40] More than this, it detailed that "the merchant galleons coming to Egypt and other major cities, the small and large *caïques* travelling among the islands, and the ships of the protected foreigners in the waters of my Sublime State, are [to receive] the endeavor and exertion of protection and security from the snares and damages of pirates."[41] As Gillis reminds us, coasts are where the majority of fishing and shipping takes place, where populations benefit from the rich ecological systems produced by the shore's ecotone.[42] This is where the spatial distinctions that defined the Ottoman maritime territory make some sense; in order to secure Ottoman waters, temporary assertions over the sea were occasionally necessary, all with the aim of protecting the people living on the coasts in Europe, Asia, and the islands, as well as those trading or sailing in what the Ottomans considered to be their waters, more often than not centered on the coast.

As has been demonstrated in the responses to both domestic and foreign pirates, these waters were clearly defined, and the Ottoman state was able to use both force and legal measures to assert its authority. However, for the most part, with the exception of the maritime regulations, Ottoman claims over the sea were not total, but relied on specific features, especially ports and forts.[43] Islands were different in providing clear landmarks in the open sea for Ottoman control, and bases for the exertion of that control. However, maritime territoriality was not defined by *insularité*, but rather by *littoralité*; the coasts of an island, and coasts in general, defined what separated *şu* from *baḥr*, making Crete no different in that sense from Anatolia. What is important in considering Ottoman insularities based on the state's attempt to define and enforce its rule against pirates and other maritime violence is therefore to think about islands as coastal entities with their own territorial waters. Further investigation needs to be done into Ottoman notions of authority at sea across time and space. However, it seems reasonable to conclude from this brief study that islands as a group did not provide a maritime border defining the limits of Ottoman control by virtue of being islands, but rather the coasts of each island projected its own maritime territory out to sea, for which liquid space the Ottoman state was responsible for ensuring security and peace, much as it was on land.

Acknowledgements

I would like to express my gratitude to the anonymous reviewer for their very constructive critiques that were of great help in improving this article's structure and presentation. I would also like to thank Sam Dolbee, Arianne Urus, and Güneş Işıksel for helping me talk through various aspects of Ottoman maritime territoriality: Michael Talbot et. al., "Commerce, Law, and Ottoman Maritime Space," *Ottoman History Podcast*, 2015 (http://www.ottomanhistorypodcast.com). I would also like to thank David Worthington, whose "Firths and Fjords" conference held at Dornoch, Scotland in 2016 proved to be extremely influential on my thoughts on this subject, in no small measure through his questions and encouragement Last, but certainly not least, I am very grateful to Antonis Hadjikyriacou for inviting me to contribute to this project, and for his advice and inspiration throughout.

Notes

1. The work of Colin Heywood has been particularly important for me in framing an understanding of Ottoman maritime territoriality, especially, Colin Heywood, "Ottoman Territoriality Versus Maritime Usage: The Ottoman Islands and English Privateering in the Wars with France (1689–1714)," in *Insularités ottomanes*, Nicolas Vatin and Gilles Veinstein, eds. (Paris: Maisonneuve & Larose, 2004), 145–176. More generally, see Robert David Sack, *Human Territoriality: Its Theory and History* (Cambridge: Cambridge University Press, 1986).
2. Başbakanlık Osmanlı Arşivleri (Prime Ministry's Ottoman Archives, BOA), C.BH 16/794, beginning of Şaban 1183 (end of November 1769).
3. Ibid. For an assessment of Ottoman security measures in the Levant, see Yusuf Alperen Aydın, "18. yüzyılda Osmanlı Devleti'nin Ege (Adalar) Denizi Doğu Akdeniz'e yönelik güvenik parametreleri," *Osmanlı Araştırmaları / The Journal of Ottoman Studies* 45 (2015): 161–184.
4. My ongoing research into Ottoman efforts to protect their Mediterranean waters charts regular and significant naval operations that used both corsairing and anti-piracy missions between the 1730s and 70s, and demonstrates the importance to the Ottoman state of ensuring maritime protection and security in the eighteenth century.
5. Dina Rizk Khoury, "Violence and Spatial Politics between Local and Imperial: Baghdad, 1778-1810," in *The Spaces of the Modern City: Imaginaries, Politics, and Everyday Life*, eds. Gyan Prakash and Kevin Kruse (Princeton: Princeton University Press, 2008), 181–213.
6. Nicholas Vatin, "Un Territoire 'bien gardé' du sultan? Les Ottomans dans leur *vilâyet* de Basra, 1565-1568," in *The Ottoman Middle East: Studies in Honor of Amnon Cohen*, eds. Eyal Ginio and Elie Podeh (Leiden: Brill, 2014), 90.

7. I have given some initial thoughts about these maritime regulations, and continue to research their implications for Ottoman maritime territoriality and foreign relations. See Michael Talbot, "Ottoman Seas and British Privateers: Defining Maritime Territoriality in the Eighteenth-Century Levant," in *Well-Connected Domains: Towards and Entangled Ottoman History*, eds. Pascal Firges, Tobias Graf, Christian Roth, and Gülay Tulasoğlu, (Leiden: Brill, 2014), 54–70.

8. BOA C.HR 118/5877, middle of Safer 1193 (beginning of March 1779). *"Düvel-i Avrupa miyānlerinin ḫuṣūmet ve şıḳāḳ taḥaddus ve sābıḳü'l-beyān şurūt-u deryā murāʿātda düvel-i merḳūme gemilerinden baʿżı gūne tehāvün ve ḳuvʿata bināʾen nehiyet ḥudūd-u İslāmiye olan memleket-i Moraʾnıñ ve velāsından cezīre-i Giridʾe kıldınca bir ḫaṭṭ farż olunub."* A subsequent command in November clarified that the borders would go from the Morea, to Crete, to the extreme western borders of Egypt.

9. Domenico Alberto Azuni, *Origine et progrès du droit et de la législation maritime, avec des observations sur le consulat de la mer* (Paris: Cérioux, 1810), 227.

10. Talbot, "Ottoman Seas," 65–66.

11. BOA C.HR 140/6968, end of Muharrem 1157 (middle of March 1744).

12. Certainly the term *cezīre* could be taken as a form of peninsula, much as the Arabian Peninsula is often referred to as *al-jazīrah al-ʿarabiyyah*. However, that the Ottomans occasionally considered Morea to be an island, or at least island-like, and that Morea could be said to share a number of features with other large islands, means that it would be extremely interesting to examine Morea within the wider framework of insularity. If we are to understand insularity as being more than "a body of land surrounded by a body of water," then the insularity of peninsulas must be taken into consideration. For an interesting framing of such an analysis, in the same place but a rather different time, see Christy Constantakopolou, *The Dance of the Islands: Insularity, Networks, the Athenian Empire, and the Aegean World* (Oxford: Oxford University Press, 2007), especially chapters 1, 4, and 5.

13. C.HR 121/6030, end of Şaban 1157 (beginning of October 1744).

14. Palmira Brummett, *Ottoman Seapower and Levantine Diplomacy in the Age of Discovery* (Albany NY: State University of New York Press, 1994), 13. Examples of the use of "fluid" in describing borders/frontiers include, Gülrü Necipoğlu, "Connectivity, Mobility, and Mediterranean 'Portable Archaeology:' Pashas from the Dalmatian Hinterland as Cultural Mediators," in *Dalmatia and the Mediterranean: Portable Archaeology and the Poetics of Influence*, ed. Alina Payne (Leiden: Brill, 2014), 313; Janet Klein, "State, Tribe, Dynasty, and the Contest over Diyarbekir at the Turn of the 20th Century," in *Social Relations in Ottoman Diyarbekir, 1870-1915*, eds. Joost Jongerden and Jelle Verheij, (Leiden: Brill, 2012), 153; Cem Emrence,

Remapping the Ottoman Middle East: Modernity, Imperial Bureaucracy and the Islamic State (London and New York: I.B. Tauris, 2011), 92.

15. An excellent critique of "fluidity" was given by Antonis Hadjikyriacou and Daphne Lappa in their paper entitled "Exploring the Conceptual Boundaries of the Concept of Fluidity: Early Modern Contact Zones in the Adriatic and the Eastern Mediterranean," at the International Conference *Well-Connected Domains: Intersections of Asia and Europe in the Ottoman Empire*, University of Heidelberg, November 10–12, 2011.

16. This was achieved by searching for the term "fluid borders" (case insensitive, 1800–2014) into the *Google Books Ngram Viewer*, http://books.google.com/ngrams. For example, see Vangelis Calotychos, *The Balkan Prospect: Identity, Culture, and Politics in Greece after 1989* (New York and Basingstoke, 2013), 3–5. An insightful discussion of the idea of the border in this period can be found in Mathijs Pelkmans, *Defining the Border: Identity, Religion, and Modernity in the Republic of Georgia* (Ithaca: Cornell University Press, 2006), 1–16.

17. Rifa'at Abou-el-Haj, "The Formal Closure of the Ottoman Frontier in Europe, 1699–1703," *Journal of the American Oriental Society* 89, no. 3 (1969): 475; Reşat Kasaba, *A Moveable Empire: Ottoman Nomads, Migrants, and Refugees* (Seattle: University of Washington Press, 2009), 58–59. Virginia Aksan's *An Ottoman Statesman in War and Peace: Ahmed Resmi Efendi, 1700-1783* (Leiden: Brill, 1995), along with her article "Ottoman Political Writing, 1768–1808," *International Journal of Middle East Studies* 25, no. 1 (1993): 53–69, remain two of the most important works in understanding the impact of this diplomatic change on the Ottoman governing elite.

18. Gábor Ágoston, "Where Environmental and Frontier Studies Meet: Rivers, Forests, Marshes, and Fortifications along the Ottoman-Hapsburg Frontier in Hungary," in *Frontiers of the Ottoman World*, A.C.S. Peacock, ed. (Oxford: Oxford University Press, 2009), 81–94.

19. David C. Kang, *East Asia Before the West: Five Centuries of Trade and Tribute* (New York: Columbia University Press, 2010), 139.

20. Dina Rizk Khoury, *State and Provincial Society in the Ottoman Empire: Mosul, 1540–1834* (Cambridge: Cambridge University Press, 1997); Colin Heywood, "The Frontier in Ottoman History: Old Ideas and New Myths," in *Frontiers in Question: Eurasian Borderlands, 700–1700*, eds. Daniel Power and Naomi Standen, (London: Palgrave Macmillan, 1999), 228–250; Gábor Ágoston, "A Flexible Empire: Authority and its Limits on the Ottoman Frontiers," *International Journal of Turkish Studies* 9 (2003): 15–31.

21. Hugh R. Clark, "Frontier Discourse and China's Maritime Frontier: China's Frontiers and the Encounter with the Sea through Early Imperial History," *Journal of World History* 20: 1 (2009): 1–4.

22. Ibid., 20
23. Michel Fontenay, "Les Îles de la Méditerranée occidentale: un contre-portrait de l'insularité ottomane?," in *Insularités ottomanes*, eds. Nicolas Vatin and Gilles Veinstein, 23.
24. A useful resource for the discussions in coastal history is *The Coastal History Blog*, hosted by the Port Towns & Urban Cultures project at the University of Portsmouth: http://porttowns.port.ac.uk/coastal-history-blog/. Two of Isaac Land's early posts on this site give an interesting introduction to the subject: "Blog 2: What Makes Coastal History Distinct? Part One of Two", 4 November 2013; and "Blog 3: What Makes Coastal History Distinct? Part Two of Two", 20 November 2013. David Worthington has also contributed to this site—"Blog 33: Firths and Fjords", 10 November 2015—and also maintains his own blogsite via the University of the Highlands and Islands, thinking about coastal history and adjacent coasts, *Firths and Fjords: Comparative Historical Perspectives on Adjacent Coasts*: https://firthsandfjords.com/ . My own contribution to the *Firths and Fjords* blog is a reflection on two outstanding conferences on island and coastal history, Worthington's "Firths and Fjords" (University of the Highlands and Islands, 31 March to 2 April 2016), and Antonis Hadjikyriacou and Sakis Gekas' "Insularities Connected" (Institute for Mediterranean Studies, FO.R.T.H., 10–12 June 2016), both of which have been instrumental in shaping my ideas on these subjects: Michael Talbot, "Insularities and Littoralities Connected: New Trends in Research from the Moray Firth to the Mediterranean," 14 June 2016.
25. Isaac Land, "Review Essay: Tidal Waves: The New Coastal History," *Journal of Social History* 40:3 (2007), 731–743 at 740.
26. David Harvey, *Justice, Nature & the Geography of Difference* (Oxford: Blackwell, 1986), 261.
27. John Gillis, "Not Continents in Miniature: Islands as Ecotones," *Island Studies Journal* 9:1 (2014), 155–166 at 163–4.
28. Ibid., 164.
29. BOA C.BH 254/11749, 1 Muharrem 1172 (5 September 1758).
30. Cyprian Broodbank, *An Island Archaeology of the Early Cyclades* (Cambridge: Cambridge University Press, 2000), 40–41. My thanks go to Antonis Hadjikyriacou for pointing out this fascinating study.
31. BOA C.BH 91/4391, beginning of Rebiülevvel 1122 (beginning of May 1710).
32. For other definitions and implications of "open seas," see Molly Greene, "The Ottomans in the Mediterranean," in *The Early Modern Ottomans: Remapping the Empire*, eds. Virginia Aksan and Daniel Goffman (Cambridge: Cambridge University Press, 2007), 104–116 (especially 111–116).

33. The following are some of the studies that have been most influential in shaping my thoughts, rather than an attempt to produce a detailed summary of works on Ottomans and the sea: Vatin and Veinstein, eds., *Insularités ottomanes*; Molly Greene, *Christians and Muslims in the Early Modern Mediterranean* (Princeton: Princeton University Press, 2000); Palmira Brummett, *Ottoman Seapower*; Edhem Eldem, *French Trade in Istanbul in the Eighteenth Century* (Leiden: Brill, 1999); Maria Fusaro, Colin Heywood, and Mohamed-Salah Omri, eds., *Trade and Cultural Exchange in the Early Modern Mediterranean: Braudel's Maritime Legacy* (London and New York: I.B. Tauris, 2010); İdris Bostan, "Osmanlılarda deniz sınırı ve karasuları meselesi," in *Türkler ve Deniz*, ed. Özlem Kumrular, (İstanbul: Kitap Yayınevi, 2007), 47–62; Edhem Eldem, "Kontrolü kaybetmek: 18 yüzyılın ikinci yarısında doğu Akdeniz'de Osmanlı varlığı," in *Türkler ve Deniz*, ed. Kumrular, 63–78; Edhem Eldem, "Strangers in their Own Seas? The Ottomans in the Eastern Mediterranean Basin in the Second Half of the Eighteenth Century," *Studi Settencenteschi* 29/30 (2010): 25–58; Elias Kolovos, "Insularity and Island Society in the Ottoman Context: The Case of the Aegean Island of Andros (Sixteenth to Eighteenth Centuries)," *Turcica* 39 (2007): 49–122.

34. The discussions over *mare liberum* and *mare clausum* in the seventeenth century set the principles for much of our understanding of maritime territoriality, in particular Hugo Grotius's assertion that "what cannot be occupied cannot be the property of anyone, for all property has arisen from occupation." For a brief discussion of these debates, see Talbot, "Ottoman Seas," 55–57. On Ottoman authority over rivers, a subject that requires more attention, see Ayşe Kayapınar, "Les Îles ottomanes du Danube au XVIe siècle," in *Insularités ottomanes*, eds. Nicolas Vatin and Gilles Veinstein, 177–202.

35. BOA, C.BH 13/627, middle of Şeval 1179 (end of March 1766).

36. Brummett, *Ottoman Seapower*, 12.

37. There is an extensive body of literature of defining the land frontiers, particularly in the West: Pál Fodor and Géza Dávid, eds., *Ottomans, Hungarians, and Habsburgs in Central Europe: The Military Confines in the Era of Ottoman Conquest* (Leiden: Brill, 2000); Mark Stein, *Guarding the Frontier: Ottoman Border Forts and Garrisons in Europe* (London and New York: I.B. Tauris, 2007); Virginia Aksan, *Ottoman Wars, 1700–1870: An Empire Besieged* (London and New York: Longman, 2007); Plamen Mitev, Ivan Parvev, Maria Baramova, and Vania Racheva, eds., *Empires and Peninsulas: Southeastern Europe between Karlowitz and the Peace of Adrianople, 1699–1829* (Münster: LIT Verlag, 2010).

38. BOA, HAT 241/13504, undated (1791/2).

39. Gilles Veinstein, "Le Législateur ottoman face à l'insularité: L'enseignement des *Kânûnnâme*," in *Insularités ottomanes*, eds. Nicolas Vatin and Gilles

Veinstein, 91. *"Nous tentons...à discerner quelle place et quel rôle propres cet État réserve à ces possessions d'un genre particulier qui les distingue de toutes ses autres provinces, elles-mêmes de natures diverses."*

40. BOA C.BH 91/4391.
41. Ibid. *Mışır ve emṣār-ı sāʾireye ẕehāb ve iyāb eden tüccār ḳalyonlarını ve aḍalar aralarında gezen ṣaġīr ve kebīr ḳayıḳları ve devlet-i ʿaliyem ṣularında müsteʾmin sefīnelerini keyd ve mużırrın ḳorsandan viḳāye ve muḥāfaẓaya ṣarf-ı himmet ve iḳdam.*
42. Gillis, "Not Continents," 164.
43. For a consideration of forts as markers of power, see Palmira Brummett, "The Fortress: Defining and Mapping the Ottoman Frontier in the Sixteenth and Seventeenth Centuries," in *Frontiers*, ed. Peakcock, 31–56. More comparative work needs to be done on Ottoman coastal fortifications, on the mainland, and in the islands.

Maritime Warfare in the Aegean and Ionian Islandscapes: Safai's History of the 1499 Lepanto Expedition

MURAT CEM MENGÜÇ

Introduction

In 1512, a Mevlevi sheikh named Safai (d. 1521) wrote a history of the 1499 Ottoman expedition to the Venetian colonies in Lepanto (present-day Nafpaktos). Entitled *Fethname-i İnebahtı ve Modon* (The Book of the Conquest of Lepanto and Modon [Methoni]), the book was written in verse, but in Turkish vernacular. This was in sharp contrast to the more sophisticated language of contemporary court histories, and one of the reasons why this source is interesting. Upon its completion, Safai's *Fethname* was presented to Sultan Bayezid II (r. 1481–1512). Nothing is known of the book's readership, the extent of its circulation, or the number of copies produced. As the earliest known Ottoman maritime history book, the *Fethname* focuses on one specific expedition, describing naval activities and life on board in detail. According to Safai, the *Fethname* is based on his own personal experiences or those of sailors he had met during and after the expedition. This precise feature also sets the *Fethname* apart from other similar histories, since it is rare to find an eyewitness account in early Ottoman history books, let alone among the genre of *fethname*s. Nevertheless, subsequent generations of historians appear to have ignored Safai's text: his name was not cited in any later histories, and was only briefly mentioned in a near contemporary biographical dictionary.[1]

Here, I focus on some of the most dominant themes of the *Fethname*, examining how they are related to the Aegean and Ionian islandscapes. In doing so, I shall provide a concise overview of the text, highlighting certain noteworthy patterns that are persistent throughout the narrative: its graphic and detailed depiction of naval battles; emotional denunciations of warfare; the vicissitudes of day-to-day life on board; and the quest for provisions and water in particular, whereby islands occupy an important, if not exclusive, role.

Safai and his Fethname

Safai was born in the town of Sinop on the Black Sea coast, historically known for its ship-building industry. In one of his rare autobiographical statements, Safai claims that he had participated in the conquest of Constantinople in 1453.[2] If true, we may assume that he was a member of the Ottoman navy, which played an important role in the operations during the siege. Safai also claims to have been present in five further expeditions without, however, providing more details.[3] Additionally, he served as the personal secretary to İskender Paşa (d. 1506), the Grand Vizier of Sultan Bayezid II (r. 1481–1512).[4] He was also known as the personal secretary of the sheikh of the *Galata Mevlevihanesi* (The Mevlevi Convent of Galata), which was founded by İskender Paşa.[5] Finally, Safai was involved in spiritual pursuits himself; at his old age he ran his own convent in Galata, frequented by distinguished Ottoman captains and sailors. Indeed, these facts corroborate the author's direct and intimate relationship to maritime life.[6]

Information on the book itself is much scarcer, scattered, and sometimes conflicting. There is no definite date for the completion of the *Fethname,* as it lacks a traditional colophon. In light of the library seal of Bayezid II on the first and last folios as well as the numerous praises to this sultan, the year 1512 could be considered as *terminus ante quem*.[7] Franz Babinger claims that Safai composed two texts (*Gazavat-ı Bahriyye* and *Fethname-i İnebahtı ve Modon*) neither of which was located at the time of his research.[8] Evidently, the latter text has become available since then, and remains the only known copy to this day.

In contrast to Franz Babiger, Safai's contemporary Sehi Beg states that Safai wrote only one book. Though it does not mention its title, he states that it dealt with the activities of the famous Ottoman navy commander Kemal Reis (c. 1451–1511) who served during the expedition to Lepanto.[9] In all probability, the *Fethname* is the book Sehi Beg referred to, since a good part of it is dedicated to Kemal Reis. More interestingly, the *Fethname* comes across as a literary response to the criticism waged against Kemal Reis' conduct during the expedition. Safai clearly states that there were unjustified accusations against some captains, mainly Kemal Reis, and that some people lied and gossiped about the nature of the events during the expedition. Such occurrences made it necessary for him to inform the Sultan about the truth, he argues.[10] In this respect, the *Fethname* comes across as an apologia on behalf of the accused Ottoman sailors, aiming to salvage the reputation of, among others, the disgraced Kemal Reis.

On another level, there is the question of where narrative agency lies. Safai writes in first person, and states that he was conveying the information related to him by eyewitnesses he had interviewed. However, he also narrates how he was recruited by Kemal Reis to serve during the expedition.[11] By any calculation, he must have been nearly 90 years old at the time of his recruitment for the expedition. This casts serious doubts on the validity of this piece of information. It is, therefore, more plausible to assume that Safai's narrative is based less on his own experiences on board, and more the accounts he collected from sailors who visited him at his convent. The chronology of the book also supports this interpretation. The narrative is non-linear, lacks consistency, and it is regularly interrupted, while events previously described are revisited from a different vantage point. This structure suggests that the book was not conceived as a uniform and singular narrative, but as a collection of distinct narratives, all related to the same set of events. In other words, it mimics the author's encounters and interviews with different sources, one after the other. Had Safai personally attended the expedition participated in the daily events, this could not have been the case.

Maritime Warfare in the *Fethname*

Commonly, all *fethname* texts open with long eulogies to God, Prophet Mohammad, and the early caliphs, before praising the incumbent Sultan. Safai follows this exact format. More than anything else, the *Fethname* is a book about a specific maritime expedition, and its chapter titles indicate a narrative organized according to specific battles and/or main naval commanders, e.g., "*Ceng-i Barak*" (The battle of Barak [Reis]),[12] "*Vusul-u Ahmed Paşa ve Ceng-i Hüvelmiç*" (Ahmed Paşa's arrival and the battle of Chlemoutsi).[13] The earliest historical chapter appears in the thirty-fourth folio of the manuscript, and is titled "*Tezyîn-i Küştehâ-ı Kosdandiniyye ve Sebeb-i Tâlif*" (Decoration of Constantinople's Ships and the Purpose of Writing [this book]). This chapter describes the preparations of the navy before the 1499 Ottoman expedition against Venetian forts in the Aegean. It provides information about how Sultan Bayezid II assigned Davud Paşa as the commander in chief of the navy, and Haydar Kaptan was put in charge of supervising the preparations. He lists the fleet's military capability, describing its ships, weaponry, crew, and cargo. But most importantly, he explains to his readers why he wrote the *Fethname*:

> I have been to the Balkans many times / six times I joined the expeditions to the islands
> By the grace of God I arrived at Modon / I saw the conquest of Lepanto
> Being alive, it then occurred to me / I should also send my words to an expedition
> I should explain to you [Sultan] what happened / I should clarify to you the situation
> Because during this expedition I saw / so much blood being spilled in the name of *cihad*
> I saw decapitated heads, used like cannon balls / I saw blood, run like river into the sea
> I saw many beauties die because of the sword / I saw many boats ruined because of fire
> Those who tried to escape drowned in the sea / a thousand lives weren't worth a straw

> Some fell dead, their private parts were open / cannons hit heads, and destroyed them
> Some hid themselves in the hull, they were scared / some hid themselves behind barrels
> They said we did not know this could be / if we did we would not have boarded this boat.[14]

As this extract makes clear, Safai wishes to be brutally explicit about the horrors of maritime warfare. Throughout the *Fethname*, he remains true to this goal. Even where there is some pleasant imagery, there are also strong undertones of sadness. Indicative is the chapter entitled "Voyage to the Islands."[15] There, Safai describes a quiet evening after the fleet had settled in the town of Eceabat (Maydos) on the southern coast of the Dardanelles. Soldiers were calm, food was shared, and the lanterns were lit. Some prayed, others talked, and in this tranquility everyone was able to listen to the sound of the *ney* (flute). In the midst of this imagery of rare serenity, Safai's eye catches a young *acemoğlanı* (a janissary foot soldier), sitting at a quiet corner on his own, and whose face became visible when he took his headpiece off, which he then rested upon his knees.[16] For Safai, this is how night falls on a boat on the eve of war: with the moody sound of *ney*, and the reflection of an innocent face who was neither an experienced sailor, nor a man who knew much about war.

Soon enough, the reality of war becomes clear. Even before exiting the Dardanelles (in the vicinity of Boğazkesen [Abydos] across from Eceabad), a small skirmish with a local Venetian colony occurs. This incident reveals that such enclaves of previously peaceful coexistence within the Ottoman domains also became zones of hostility after the declaration of war with Venice. The Ottoman navy attacks and plunders the Venetian colony, heading for the fortress of Yeni Hisar at the tip of the Dardanelles. There, they find favorable winds, which take them to İskire (Skyros) in good time.[17] Once there, Safai notes that some sailors headed for the land immediately, wanting to engage in looting and war, while others stayed on board. A new theme emerges here, according to which land is not always a better place to be. In fact, throughout the *Fethname*, embarking on the shore, as it was the case in İskire, almost always guarantees a direct encounter with the enemy. It is a more life-threatening situation than being on a ship, where sailors seem to have a better chance of escaping a hostile confrontation, if their commanders decide so.

As it was previously mentioned, Safai is particularly interested in the graphic scenes of violence. He seldom leaves out the terrible images of battlefield. These, he combines with references to *gaza* and *cihad*, i.e., holy war. Although, for Safai, holy war remains a pillar of Islamic faith, his negative depiction of war creates an almost bipolar narrative.[18] What makes his narrative stand out is the emphasis on maritime warfare and day-to-day life on board. These are topics his audiences were less likely to be familiar with, at least in such graphic detail. Non-combatant populations came to know about naval battles through narrations, or by watching them from afar (in the case of coastal populations). People were more likely to be accustomed to land-based warfare either through their own direct experience as warriors, or as non-combatant populations affected by war. Indeed, descriptions of naval warfare are more sensational, awe-inspiring, and perhaps even exotic. For instance, when a ship is destroyed, it most likely sinks in to oblivion. Bodies may never be found, a proper burial never received, and a grave stone to commemorate one's legacy is too much to ask for. Safai's apparent emphasis on the brutality of war lies on the fact that he had the rare privilege of collecting eyewitness accounts. His informants probably remembered the most dramatic scenes they had witnessed, and specifically choose to share the most breathtaking memories of their voyage. This is an important detail to keep in mind when one reads Safai: the vivid depiction of naval battles and the harsh realities of everyday maritime life through eyewitness accounts is what distinguishes his work from contemporary histories to a great extent.

One of the best examples of such vivid imaginary is found in the chapter titled "*Ceng-i Barak*" (Barak [Reis]'s Battle). According to Safai, this battle took place on 5 Muharrem 904 (12 July 1499) near Paylos (Pylos), in the vicinity of Sapientza (which the Ottomans named Barak adası after Barak Reis), off the Southwest coast of the Peloponnese, across Modon. This is the same Barak who sent to France as a secret agent in 1486 in search of Cem Sultan.[19] The battle itself later acquired an almost mythical status in the Ottoman/Turkish maritime history. In modern popular imagination Barak Reis is remembered as one of the most competent captains of the Lepanto expedition. He is believed to have committed a heroic feat during the battle by riding his boat to the center of the Venetian fleet and setting his ammunition on fire. His exploding ship, with him as the last man standing on board, sank two of the important Venetian ships belonging to the leading commanders of the enemy.[20] Although it is true that during this battle Barak Reis' boat and two Venetian boats were destroyed, Safai's rendition of the events is remarkably different.[21] Safai writes:

> Now, listen to the weirdness of things / so you will know their darkness
> On year 905 [A.H.] / they [the Ottomans] locked their claws with the infidel
> Understand the meaning of Safai's word / it was the 5th of Muharrem indeed
> He who doesn't listen to my word / won't distinguish the hero from the villain
> Everyone knew their own story / but some saw a fault in this story
> They said this or that happened / they lied about who was the real hero
> As you know, six hundred boats were there / all firing the cannon balls at once
> The sword, the boat, the rifle, all wanted to fight / cannons destroyed them when they opened their mouths
> And hear this: nobody was able / to see anything from the layers of smoke
> [Our] Arrows rained but wouldn't set them [infidels] on fire
> Thoughtfully listen to my words / understand the design of the infidels
> They were ready to fight the Turks / they were a crowded flying pack
> They fired cannons for some time / they were about five miles apart[22]
> For the time being the cannons roared / each side stood by his weapons

> Afterwards the galleons pulled their anchors / and set themselves upon the unbelievers.[23]

At this point, Safai writes, two Venetian captains attacked Barak Reis' ship, thinking that he was Kemal Reis, the commander in chief. According to Safai's informant, a Venetian captive, the two Venetian captains were Lordan (Andrea Loredan) and İstefanos (Stefanos); each one commanded a ship of three thousand barrels. Lordan's ship was taller, and he tied its anchor line to Barak's ship.[24] Safai goes on to describe the battle and how Barak Reis' ship slowly caught fire;

> It was impossible to cut it [the tie of the anchor] off / they hit each other with cannons and rifles
> From an angle poured the cannon balls / they destroyed numerous heads
> So they hit each other with arrows / [until] the boats started to look like porcupines
> They sent to the infidel their towers of light[25] / their skins started to receive scars
> Whenever one of them was cut / much blood spilled from the blade [...]
> Greased arrows reached its [Barak's] sails / setting the sails and the mast on fire
> Thus they [Ottomans] threw grenades / two of which they set on fire.[26]

At this point, Kemal Reis tried to interfere, Safai notes, and sent some smaller boats to rescue Barak Reis and his crew. But it was too late. Barak Reis' ship was already completely on fire. Everyone tried to escape the scene, according to Safai. As his words will not do justice to this incident, Safai opts for a description of an eyewitness, an *odabaşı* (a janissary officer); he was able to escape Barak Reis' ship, along with six others, most of whom were hurt during battle. Although the *odabaşı* was able to escape, when he saw Barak Reis standing alone on his burning ship, he decided to draw near him. He asked Barak Reis for his wish. Standing awestruck on his burning boat, Barak Reis was unable to utter a word,

The *odabaşı* goes on to describe how many sailors fell into the sea and drowned. An outstanding explosion followed, whereby the mast of Barak Reis's boat "shot up to the sky like an arrow."²⁷ The fire continued for the rest of the day and throughout the night, we are told.²⁸ In the end, an Ottoman and two Venetian ships were destroyed. This is one incident among many described with regard to naval battles throughout the book. Another one notes how a cannon became immobile due to inexperienced rowers near the island of Aya Mavra (Lefkada) in the Ionian Sea.²⁹ Or, there is the story of one Seyit Ali, an Ottoman captain, captured and enslaved by Venetians, containing the details of the long negotiation for his release.³⁰ However, it is not the purpose of this article to outline and describe all these incidents in detail. It would suffice to say though that they all follow the above-mentioned themes.

A Storm, a Mutiny, and the Quest for Fresh Water

War is not the only theme undercutting Safai's narrative. Another persistent topic is the quest for fresh water. While it may appear basic and trivial, Safai's recurring narration of such episodes invites the reader to evaluate an otherwise simple human necessity within the conditions of everyday maritime life. Here, coasts and islands become, literally, vital. Safai mentions the relevance of fresh water for the first time when the Ottoman fleet reaches Kızıl Esre (the red shore). This was most likely an Ottoman possession on the Northeast shore of Eğriboz (Euboea), across the Venetian stronghold of Chalkis, which the Ottomans wanted to seize. For the landing on Kızıl Esre, Safai uses the verb *sulanmak* (to be watered).³¹ The distance from the previous landing point at Skyros to Kızıl Esre is not a daunting one, roughly 30 miles. However, if this is the first occasion of replenishing fresh water reserves since the Dardanelles, then the distance is roughly 150 miles. This figure is important. Safai discusses the shortage of fresh water again when the fleet is between Benefçe (Monemvasia), from where they were unable to receive water and Koron/Modon. In each case, the distance is approximately 150 miles.³² Later on, we also read about the scarcity of water near Zakynthos. In other words, the theme of water shortage appears when the fleet was not able to replenish its provisions for roughly similar distances. Safai gives us at least a general idea of when water reserves needed to be replenished.[33]

Water becomes a recurring theme in the narrative under different circumstances. In one instance, Safai discusses how numerous boats gathered around water sources.[34] At another point, he relates how certain landings were regarded beautiful because of their water.[35] He also notes that Preveza supplied fresh water to the fleet at a time of great need.[36] Other examples include the island of Acısu (literally: bitter water; most probably Schiza),[37] or how a sailor once chased a goat, until it became thirsty enough to lead him to an otherwise disguised fresh spring.[38] This last episode, hardly to be expected in a narrative about maritime life, is indicative of how the day-to-day existence on a ship depended on the intersection between land and sea.

The most dramatic references to fresh water are encountered in the story the story of a storm, during the fleet's stay at the above-mentioned Kızıl Esre site in Euboea. According to Safai, a storm dispersed the Ottoman fleet twice from this shore. As a result, three horse ships (*at gemisi*) were lost. Two of the ships were later recovered. The third one, which possessed five large cannons, remained missing. When it arrived at an unspecified island, the locals, presumably members of a local Venetian colony, attacked it. During this battle, four of its men died, including the captain.[39] This episode most likely took place near present-day Kefireas Straight, or Kafir Boğazı (Infidel's Straight) as it was known to the Ottomans. Prior to the 1499–1500 Lepanto expedition, when diplomatic relations between the Ottomans and the Venetians were marked by a breakdown, this straight was considered a Venetian possession, hence we can speculate that this explains the name "Infidel's straight." This is the most frequented straight between Euboea and lower Peloponnese, and represents the shortest way from North to South on the Greek coast. The other straights, located between Andros and Tenos islands, and the Tenos and Mykonos islands are also famous for their strong winds.[40]

At this point, Safai's narrative ends abruptly with the conquest of Lepanto. An Ottoman council was summoned at Aspra İspitya (meaning white houses in Greek), and decided that the army should spend the winter in the Peloponnese. The fleet would continue to the second leg of the expedition, against the Venetian strongholds of Crete, Koron, Modon, and Anavarin (Navarino).[41]

In the following chapter, entitled *"Cevadet-i Cezayir"* (The Offerings of the Islands), Safai returns to the story of the lost horse ship.[42] He informs the readers that after the victory, a certain Kara Hasan was ordered to go to a general search of the nearby islands with three captains, their galleys, and two small messenger boats (*Kayd*).[43] Probably due to lack of water supplies, they soon needed to replenish their provisions. They sailed from island to island in search of fresh water, only to find the lost horse ship at their third stop. After the killing of the horse ship's captain during the encounter with the Venetian guard, a sailor named Kara Durmuş staged a mutiny, taking possession of the ship. When the ship was discovered by the expedition team, Kara Durmuş had reportedly escaped. Nonetheless, due to several inconsistencies in the sailors' stories, the ship and the crew were removed from the expedition and were escorted to Livatya (present-day Bandırma) in the Marmara Sea.[44]

Afterwards, Kara Hasan, and presumably his entourage and some of the other captains, travelled back to the shores of the Peloponnese. They first passed Çura (i.e. Çuka, Kythera), before reaching Benefçe (Monemvasia). There, they encountered the rest of the Ottoman fleet, led by Solak Reis and Kemal Reis.[45] Safai notes an important detail here, that these captains left Benefçe in search of fresh water; prior to the 1499 Lepanto expedition, Benefçe was the most important (if not sole) Ottoman possession on the Peloponnesian shores. According to Safai, it lacked enough fresh water to supply a large Ottoman fleet. For this reason, fresh water was acquired from Preveza at a later stage, from where 10,000 barrels of fresh water were supplied for the fleet.[46]

Following the lead of Solak Reis and Kemal Reis, the entire fleet headed for Koron, navigating more hostile shores. Soon, Safai writes, things took a turn for the worse:

> A strange thing happened to us / a weirdness fell upon us
> While we traveled many shores for water / we finally reached a landing
> But there we faced a problem / there were too many boats around us
> So we decided to take anchor / listen what the wind did to us!
> A storm came, we were caught / it blew from the land, so much water [on us]

> On the shore the hay was being collected / and the people there set the hay on fire
> The fire was all over us / choking us with its smoke
> Each boat [crew] had to hide within its hull / so they won't be burned by the fire
> Anyone who peeked out to see / saw birds being burned and falling.[47]

Following this incident, the fleet could no longer entertain the idea of anchoring anywhere; they sailed throughout the night, until they reached the vicinity of Koron in the morning.[48] Exhausted by this point, some men immediately went to land upon arriving at the shores near Koron. Yet again, instead of fresh water, they were confronted by Venetian troops, which dispersed after some skirmish. The Ottoman sailors continued their pursuit of provisions, finding some gardens and orchards and enjoying eating the fruits.[49] Yet, there is still no mention of fresh water. In this seemingly deserted place, they came across a monk, living alone in a small church. The Ottoman sailors killed him simply because he was a *kafir* (infidel). They then set the church on fire, along with his collection of books.

Interestingly, Safai introduces the above-mentioned incident as *bir güzel kıssa* (a beautiful story), concluding it with the statement "let us now move to something that could make you happy."[50] This refers to the capture of a Venetian informer, who disclosed the location of a hidden Venetian fleet, near Modon, behind the island of Acısu (in all probability, Schiza).[51] Hereafter, and until the end of the battle of Modon, for some ten folios, water shortage is not mentioned.[52] After the Ottoman victory in Modon, Safai provides an eloquent description that celebrates where they found abundant fresh water:

> A big water running sweet and beautiful / Created by God, the helper of people [...]
> Yes, it is only water, but a beautiful happiness / the great treasure, the pure compliment
> So they [sailors] went ashore to get water / To quench themselves full and full.[53]

VIOLENCE AND LAW IN TERRAQUEOUS SPACE 99

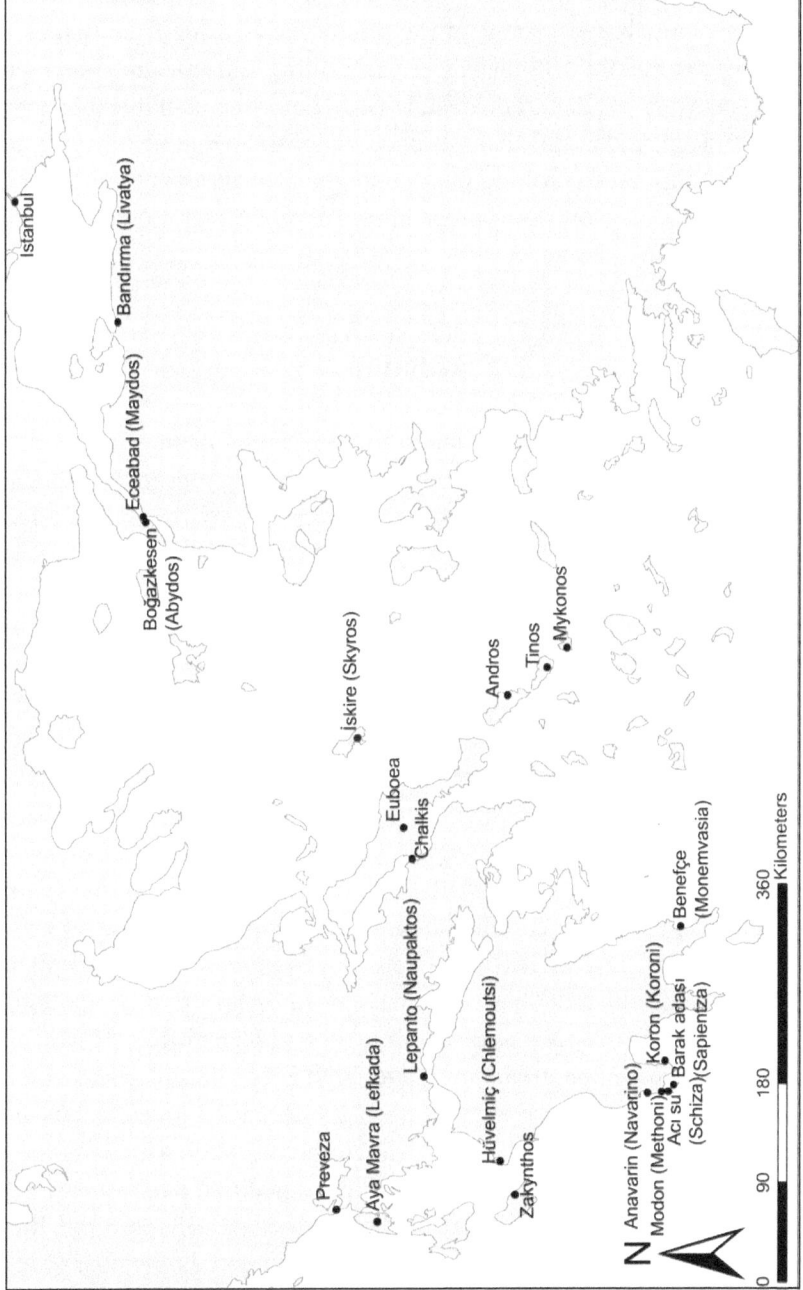

Map of the identified campaign stops mentioned in the text.

The lack of fresh water is repeatedly addressed in Safai's text, albeit in less dramatic language. He makes it abundantly clear that the absence of fresh water is the biggest problem the sailors faced, aside from battle. It is also obvious that the Ottomans were navigating a hostile and unknown territory, unaware of sources of fresh water. They either had to explore on their own or depended on the assistance of locals, while also having to reduce consumption to prolong their supplies. It is worth noting that the availability of or replenishing other kinds of provisions are completely absent from Safai's narrative.

Conclusion

In this article, I underline some of the recurring themes of Safai's *Fethname*, in order to illustrate maritime life in the Aegean and Ionian islandscape. My goal was to provide an overview of this manuscript, which has much to offer to the history of the Mediterranean at the turn of the fifteenth century. Through this overview, the role of islands in such expeditions emerges to the fore. The text itself is rich in its depictions of naval warfare, graphic and critical in terms of the battle scenes. Descriptions are enriched by numerous references to the day-to-day life of the Ottoman sailors, trying to secure provisions in a hostile environment. It is most likely the earliest Ottoman naval history book. It is also unique in terms of the information it gives regarding the first major Ottoman naval expedition against the Venetian colonies in the Aegean and Ionian seas. Not only does it offer an account of the major battles of the expedition, but it also enriches them with the stories of the lesser skirmishes, mutinies, and at least one captivity narrative. Safai's explicit statements suggest that the book may have been imagined as an apologia on behalf of wrongfully accused captains, such as Kemal Reis. However, his personal involvement in the narrative, both as an old sailor and as the person who had interviewed eyewitnesses, contributes a great deal to the uniqueness of the *Fethname*. This is also reinforced by the way Safai goes beyond the naval and military aspects of the expedition itself, providing important details on the everyday and the mundane of maritime life. There is no question that Safai was obviously an important member of the

Ottoman sailor community. His adoption of a popular genre of history writing, the conventions of which he alters by the inclusion of numerous eyewitness accounts, makes him a unique author within early Ottoman historiography.

In their essay entitled "An Outline of Ottoman Maritime History," Eyüp Özveren and Onur Yıldırım points to the "need to know the kind of people who built and manned Ottoman ships [...] Maritime history [...] is also *social* history—it has much to do with people as with the sea."[54] This criticism of twentieth-century Ottoman historiography is also applicable to that of the fifteenth century, where there is no information on the social experiences of the historical actors involved. In contrast to this trend, Safai's *Fethname* stands out, offering numerous detailed and rich eyewitness accounts. True to the genre of *fethname*, the book focuses on war. Yet, it goes beyond battle scenes and discusses other aspects of maritime life such as storms, captivity, or mutiny. In one of his famous dictums, Braudel observed that "[t]he sea is everything it is said to be: it provides unity, transport, the means of exchange and intercourse, if man is prepared to make an effort and pay a price. But it has also been the great divider, the obstacle that had to be overcome."[55] Such eyewitness accounts vividly reconstruct what one had to struggle with in that ever fluctuating liquid geography.

Notes

1. Sehi Beg (d. 955/1558–59) in his *Tezkire* (c.1538–39) states that Safai composed a book on the activities of Kemal Reis (c.1500), a famous Ottoman navy commander at the time who participated in the Lepanto expedition. However, Sehi Beg does not mention *Fethname* by name. Meanwhile, the sections of *Fethname* dedicated to Kemal Reis' activities constitute only a portion of the book. Sehi Bey, "Heşt Bihişt," in *Heşt Bihişt the Tezkire by Sehi Beg*, ed. Günay Kut (Cambridge: Harvard University Press, 1978), 139.
2. Safai, *Fethnamei İnebahtı ve Modon*, No. 1271, vol. 131, f.21a, Topkapı Revan Kütüphanesi, Istanbul.
3. Ibid.
4. For more on İskender Paşa's career, see Kemalpaşazade, *Tevarihi Ali Osman* vol. 7, Şerafettin Turan ed. (Ankara, Türk Tarih Kurumu, 1997), 436–437, 473–74.

5. It is also said that İskender Paşa prevented an assassination attempt of Bayezid II in 898/1492–93. *Anonim Tevârih-i Âli Osman* (İstanbul: Türk Dünyası Araştırmaları Vakfı, 2000), 137.
6. Kut, ed., *Heşt Bihişt*, 139.
7. Safai, *Fethname*, f.1a and f.131b.
8. Franz Babinger, *Die Geschichtsschreiber der Osmanen und ihre Werke* (Leipzig: Otto Harrassowitz, 1927); Franz Babinger, *Osmanlı Tarih Yazarları ve Eserleri*, trans. Coşkun Üçok (Ankara: Kültür Bakanlığı, 1992), 55-6.
9. *Heşt Bihişt the Tezkire by Sehi Beg*, 139.
10. This is best illustrated in a long explanation of why Kemal Reis was unable to engage with the enemy during a major battle. Safai, *Fethname*, f. 59a.
11. Safai, *Fethname*, f.20b.
12. Safai, *Fethname*, f.34b.
13. Safai, *Fethname*, f.49b.
14. "[N]ice kez ruma geldim ve gitdim / ciddeye altı kez sefer etdim / hamdullah Moton a hem erdim / İğribuzun fethini gördüm / hatırıma bu geldi kim hayatta / sefer idem sefer diyem yazam bu sözleri ta / vafığın hoş sana beyan edeyin / hub vazıştın (vaziyyetin) sana ayan ideyin / görmişem ben bu seferde zira / nice kanlar döküldü oldu gaza / başlar top oluban atıldı / kan revan olundu bahre katıldı / düşdü kılıç yüzünde merveşler / yakdı nice gemi ateşler / dökülür bahre boğulur kaçan / bir çöpe geçmez idi bin insan / kimi düştü ayabı açıldı / başa taş irdi bin saçıldı / kimi korkudan ambara indi / kimi baya ardına irdi / kimi didi biz bu hal bilmezidik / yoksa biz bu kukaya girmez idik." Safai, *Fethname*, f. 20b–21a.
15. Safai, *Fethname*, f.20b.
16. Safai, *Fethname*, f.25a
17. Safai, *Fethname*, f.27a.
18. Safai, *Fethname*, f.12a.
19. Victor L. Ménage, "The Mission of an Ottoman Secret Agent in France in 1486," *The Journal of the Royal Asiatic Society of Great Britain and Ireland*, 3, no. 4 (1965), 112-132; Serafettin Turan, "Barak Reis'in, şehzade Cem mes'elesiyle ilgili olarak Savoie'ya gönderilmesi," *Belleten*, 26. no. 103 (1962): 539–55.
20. A good example of this mythological depiction of Barak Reis in the popular Turkish literature is found online in the Turkish Wikipedia entry regarding Barak Reis. https://tr.wikipedia.org/wiki/Barak_Reis. Accessed August 14, 2016.
21. Piri Reis, *Kitab-i Bahriye*, W.658 *Book On Navigation*, image 282. Baltimore, The Digital Walters, The Walters Art Museum. http://www.

thedigitalwalters.org/Data/WaltersManuscripts/W658/data/W658/sap/W658 _000282_sap.jpg. Accessed August 14, 2016.

22. In Safai's measurement, one mile is roughly 0.0125 miles. Thus, the distance mentioned here would be roughly 0.625 miles. Safai, *Fethname*, f.21b.

23. "*işit şimdi işin a'cāsını / tā bilursin nedir garāsını / çünki irdik dokuz yüz pençe / kāfir ile tutuşdular pençe / nedir anla sözün safainin / beş gününde muharrem ayının / ol kişi kim sözümü işitmez / kötü hem bahadırı seçmez / her kişi bildi kendi hissesini / buldü amma bu kıssa gussasını / dil ile kim bunu beyan eyler / kunhi budur diyen yalān söyler / altıyüz kim gemi işitdin sen / top taşı atardı bir kezden / zarbı zan hem pirengū şayka tüfek / ağzın açıb ederdi daşlar fek / bir yerin kişi işit kim sen / dembedem görmezidi tutundan / ok yağar üzerine durmazdı / yakmağa kimse kādir olmazdı / sözümü fikrin ile hub anla / kāfirin fiğlini işit dinle / Türk ile cenge ola hazırlar / birbirine dolu uçtu kāfirler / attılar top eyleyib ta'cīl(?) / vardır ancak tafāvatı beş mil / toplar bir zamān kütürdeşdi / her biri yarağına duruşdu / kukalar çalışıb demir aldı / önce müminler üstine saldı.*" Safai, *Fethname*, f. 44a–44b.

24. Safai, *Fethname*, f. 44b.

25. "Towers of light" is most likely a figure of speech, evoking shots of fire.

26. "*[M]ümkün olmazdı kim idi ladifek / vurdular birbirine top ve tüfek / gecelerden ki döktü taşı / pāre pāre etdiler nice bāşı / biri birine ki vurdular sehmi / kirpiye tuttu kukalar cismi / saldılar kāfire menāreleri / vurdı tenlerine yaraları / birine kesme erse leşkerden / kān saçılır nice ki neşterden ... neft oku yelkenine erişdi / direği yelkenile tutuştu / akıbet kumbaralar atdılar / ikisin bile uda yazdılar.*" Safai, *Fethname*, f. 45a.

27. Safai, *Fethname*, f.46b.

28. Safai, *Fethname*, f.46b.

29. Safai, *Fethname*, f.94b.

30. Safai, *Fethname*, f.67a– f.71a.

31. Safai, *Fethname*, f.27b.

32. See endnote 22, and, Safai, *Fethname*, f.21b.

33. John Guilmartin, *Gunpowder and Galleys: Changing Technology and Mediterranean Warfare at Sea in the Sixteenth Century.* (Cambridge: Cambridge University Press, 1974); John H. Pryor, *Geography, Technology and War: Studies in the Maritime History of the Mediterranean, 649–1571.* (Cambridge: Cambridge University Press, 1988), 165–192.

34. Safai, *Fethname*, f.32a.

35. Safai, *Fethname*, f.51b.

36. Safai, *Fethname*, f.98b.

37. Safai, *Fethname*, f.36b.

38. Safai, *Fethname*, f.39a.
39. Safai writes that Mustafa Bey was put in charge of taking this boat back to *Livatya* (modern Bandırma), a major Ottoman shipyard city on the Southern coast of Marmara. See ibid,. f.28a-b, and f.30b.
40. Christy Constantakopoulou, *The Dance of the Islands: Insularity, Networks, the Athenian Empire, and the Aegean World* (Oxford/New York: Oxford University Press, 2007), 25.
41. Safai, *Fethname*, f.29a.
42. Safai, *Fethname*, f.29b.
43. Safai, *Fethname*, f.30b.
44. Safai, *Fethname*, f.31a.
45. Safai, *Fethname*, f.31b.
46. Safai, *Fethname*, f.98b.
47. "[B]ir acayib vakıa erdi bize / bir gariğbaş ne turfa geldi bize / su için nice merse dolandık / bir yaka dibine demir saldık / uğradık anda bir katıyıyla / çok gemi var bizimle bile / demir almağa bir mecal olduk / yel ne kıldı işit ne hal olduk / fırtına oldu ceyrana kaldık / kurudan esti sudan usandık / ol yakaya bil od birikmiş imiş / onu geçenler meğer ki yakmış imiş / üstümüze saldı ol odu / boğar yazdı bizi anun devidu (dumanı) / yaptılar her gemide anbarı / yanmaya ta güherc ile varı / kim o dem gün yüzün görürdü / kuşı bir yan ider düşürürdü." Safai, *Fethname*, f.32a.
48. Safai, *Fethname*, f.32b.
49. Ibid.
50. Safai, *Fethname*, f.33a.
51. Ibid.
52. Safai, *Fethname*, f.39a.
53. "[B]ir büyük su akar güzel şirin / hoş yaratmış onu huda'ı mü'in ... sudur amma güzel saadetdur / hoş ganimet ki pür letāfetdur / çıkdılar daşra kim sulanmağa / içiben kana kana kanmağa." Safai, *Fethname*, f.42a.
54. Eyüp Özveren and and Onur Yıldırım, "An Outline of Ottoman Maritime History," in *New Directions in Mediterranean Maritime History*, Gelina Harlaftis and eds. Carmel Vassallo, (St John's, Newfoundland: International Maritime History Association, 2004), 151.
55. Fernand Braudel, *The Mediterranean and the Mediterranean World in the Age of Philip II* (New York, Harper & Row Publishers, 1972), 276.

PART III

Regulating Islands

Blockading an Island: Collective Punishment, Islanders, and the State in the "Largest" Island at the End of the Nineteenth Century

FATMA ŞİMŞEK

In April 1892, a wheat-laden ship with the Greek flag was left stranded by its crew and captain near the island of Kastellorizo (Megisti, literally: "largest;" Tk: Meyis) because of strong currents and bad weather.[1] The ship was soon looted by the islanders, who unloaded the cargo on their boats carrying it ashore. As a result, the French insuring firms demanded restitution of the looted cargo from the Ottoman state.

Incidents of looting are commonly found in the Ottoman archives. This case, however, stands out as it escalated to a complex entanglement between the Ottoman Government, the people of Kastellorizo and the French Embassy, with diplomatic, political, social and economic dimensions. This is a rare instance when we can document the procedure of collecting compensation for looting, as well as observe the language and methods the Ottoman state used in dealing with the affair. In this instance, and while responding to the French insurance companies' demand for restitution, the Ottoman state enforced a blockade as collective punishment to the people of Kastellorizo, thus isolating the island from the outside world. Scrutinizing the relevant documentation sheds light upon important aspects of center-province relations. One particularly intriguing facet concern the opinions on the islanders expressed by centrally-dispatched officials who were appointed to deal with the situation.

Within this context, thinking about location requires a consideration of geography, the lifestyle of the islanders, their relations with each other, with other societies. Simply put, location and geography are among the primary components which form a society's identity, political choices and tendencies. In particular, the way relations of power are articulated in an insular setting is one of the primary concerns of this article.

There was no uniform way that the Ottoman state dealt with its islands. Factors such as the size of an island, its geopolitical importance, income resources, historical conditions or distance from the mainland are very important in terms of establishing administrative mechanisms and maintaining them. For example, it was difficult to establish an effective administrative mechanism if surplus-extraction in the form of taxation was limited due to scarce resources—especially when sustenance was at stake.[2] In addition, due to the fluctuating geopolitical importance over time, these control mechanisms could be correspondingly be strengthened or weakened.[3]

Kastellorizo lies off the southwest Anatolian coast, less than two km from Kaş and 125 from Rhodes. With an area of just 12 square km, the island is characterized by a shortage of resources. It should thus be thought of as radically different to other bigger islands that the Ottoman ruled, like Crete, Cyprus or Rhodes. Kastellorizo became part of the Ottoman Empire in 1522 after the conquest of Rhodes, to which it was an attached district.[4] Because of its a homogenous social and economic structure, the Ottomans did not encounter major challenges in establishing its own administrative and economic organization, as was the case with larger, more populous, and resource-rich islands.[5]

Because of its proximity to the Anatolian coast, expenses for the fortress troops were covered by Teke and Alaiye.[6] The island residents lived on the goods they bought from Kaş, Rhodes and Kalkan. Timber shipments was of particular importance.[7] Such was the lack of local agricultural resources that from the eighteenth century onwards some inhabitants would have farms on the Anatolian coast.[8] Other requirements in terms of materials, supplies, or specialized labor could also be supplied by near-by islands through maritime links—particularly with Rhodes.[9]

The island was only visited by foreign travelers in the eighteenth century, when people from Kaş settled there, joining the small military garrison.[10] Charles Fellows, a British traveler who visited the island in the nineteenth century, described Kastellorizo as a large emporium with all kinds of goods and money from trade on the south-west route of the Mediterranean.[11] Its natural and deep harbors greatly facilitated the development of the maritime activities.[12] Yet, even the supply of fresh water was a serious issue on this island.[13]

Following this brief overview of Kastellorizo, let us shift attention to the incident that is the subject of this article. Loaded with 475 metric tons of wheat, the ship was insured for 109,000 francs.[14] The Ottoman administration on the island was initially informed about the incident by the local Greek consular agent who, alongside three local captains, examined the ship without informing the local Ottoman authorities. In their official report, the latter underlined how they were not asked to take any measures to prevent the plundering of the ship by the Greek consular agent.[15] In fact, they claimed that they were only informed about the incident after a long period of time. They added that eight or nine officers were sent to the ship preventing the Kastellorizo people, who had surrounded the ship with their boats, from stealing the wheat from the ship. As the ship's load subsided due to the plundering, the ship was pulled into the shipyard of Kastellorizo. According to other reports, however, the plundering continued there as well, and the Kaymakam of Kastellorizo not only did not do anything to prevent the plunder,[16] but the police, gendarmerie and officers were accused of being complicit since they either accepted part of the wheat or bribes in turning a blind eye to the boats loaded with wet wheat.[17]

Conflicting narratives notwithstanding, there is no doubt that the ship was plundered by the people of Kastellorizo. Acting on behalf, and protecting the interests of the insurance companies the French Embassy in Istanbul intervened, as was the norm,[18] petitioning the Porte for the amount of 109.000 francs as their damage due to the actions of the people of Kastellorizo. The French expressed their sorrow for the event denouncing the actions of the people of Kastellorizo, adding however that the Ottoman authorities also had their share of responsibility since they did not take any measures to prevent the plundering.[19] The question of establishing the events and individual responsibilities occupied the authorities for a long time.

In doing so, Istanbul went to great lengths to question, detain, and/or arrest the Kaymakam, local officials and the islanders involved in the plunder. The task of investigating and solving the problem onsite was assigned to Ohannes Efendi, who was the Kaimakam of Crete.[20] Such a move was in fact standard procedure for the Ottoman state, which frequently assigned an official inquiry to the governor of other provinces.[21]

On May 12, 1892, the Şüray-ı Devlet Başkanlığı (Council of State) decided that a trial would determine the amount of compensation that the French were entitled. It also ordered the arrest of the Kaymakam, other officials, and anyone else involved in the plundering of the cargo.[22] Akif Bey, the governor of the Province of the Archipelago (Cezayir-i Bahr-i Sefid), of which Kastellorizo was a part, asserted that the initial actions taken for the resolution of the affair were insufficient, and explicitly expressed his concerns.[23] Indeed, assigning responsibility in such an affair was no straightforward business—it was virtually impossible to know who participated in the looting and who was not. During this whole process, French pressure for a resolution to the affair was mounting. The Ottoman center was anxious to appease the French and provide a satisfactory response in order to prevent any possible diplomatic crisis.[24]

In dealing with such a delicate situation, Akif Bey, anxious to ensure peace in his province since there was a lot of tension between imperial officials and the locals, offered a different option of dealing with the situation than with the trial. A locally appointed committee would conduct an investigation, assess the amount of compensation, and collect it from the inhabitants in the form of an extraordinary taxation.[25] The rationale was to avoid putting the vast majority of the inhabitants on trial,[26] something that would further exacerbate tensions between the people of Kastellorizo, the Greek consular agency and other Ottoman officials. This proposal was adopted by Istanbul, and an imperial decree was issued.[27]

Objecting to this measure, the people of Kastellorizo petitioned against their collective punishment in the form of a tax.[28] Several delegations were sent to Istanbul, in the form of agents or representatives that would bring attention to islanders' case, protest against the tax and demanding a trial.[29] Soon, pressure from the province intensified.

Akif Bey quickly dismissed these representations. He argued that representatives and delegation members only offered vain hopes to the people of Kastellorizo, since time was of essence. Any delay would further prolong and intensify tensions, further damaging the locals.[30] Akif Bey repeated these arguments several times, also influence the center. Almost a year after the incident, the Porte issued a document on April 12, 1893 demanding that that the representative should be found and sent back, emphasizing that "this should be stated to [him] in appropriate language in case he [also] petitions the Ministry of Interior."[31]

The hostility with which the people of Kastellorizo approached the proposals for the means of assessing and collecting the tax, as well as that they would resist such a measure, were abundantly clear when Istanbul announced that Cavid Bey, the Kaymakam of İstanköy (Kos) was allocated this role. This would prove to be a controversial choice that would further embitter the people of the island due to the iron fist with which he extracted the tax from them.[32]

Coming to Kastellorizo from another island, Cavid Bey enforced a radical measure to curb the resistance of the small community by imposing a ban on anyone coming in or going out of the island. Essentially isolating the inhabitants of Kastellorizo from the outside word, especially as the Anatolian mainland was so close, Cavid Bey severed the connections between the two coasts. In doing so, he reversed what had hitherto been the standard Ottoman policy of organically linking the island with Anatolia. This policy aimed at not only ensuring the sustainability of the island populations, but was also a means of controlling it, as we saw above with reference to its military garrison. As this was the cornerstone of center-province relations throughout the centuries, it is difficult to overestimate the effects of this measure.

As one might expect, the people of Kastellorizo protested, sending several petitions signed by their community leaders, other notables, or captains throughout March 1893. In particular, they stressed the harsh nature of the measures advocated by Cavit Bey.[33] One such petition elaborated on effect the blockade had on the island's maritime activities, the main source of income for the local economy, whereby 80 ships were immobilized in the port.[34] Foreign consulates and representations also petitioned the Porte protesting against the blockade and emphasized the damage this brought to trade and commercial relations. These included the vice-consul of Austria, the consular agent of Greece, and the French Admiral at the Archipelago.[35]

Apart from the economic argument, other objects included the indiscriminate fashion with which this punishment was applied, without even a judicial procedure that would establish the facts. In articulating this point, the petitioners utilized the familiar language of petitions, whereby appealing to the Sultan's claims to be the protector of justice, they also challenged him to justify his titles and discourse.[36] In one such example, the emphasized the damage this measure was inflicting upon them, implying that this was unjust and unworthy of a justice-loving (*adaletperver*) Sultan. They stressed how there were about eighty ships of more than five tons owned by the islanders and stationed at the port, something that would have extremely negative consequences to the local economy.[37] Furthermore, they stressed the lack of any judicial procedure and committed to accepting any legal judgment if the owner of the wheat filed a suit against them and they were to be indicted.[38]

Responding to these petitions, the Justice Ministry intervened and asked the Province of the Archipelago for further information.[39] The Governor Akif Bey quickly tried to dismiss any suspicions of violation of justice. Implicitly emphasizing political contingency, he stressed how any judicial procedures would cause further delays, tensions and confusion—sticking to his original position. He went a step further, requesting the right to delegate authority to state officials and take all necessary actions.[40] Arguing that the measures taken were neither unfair nor unnecessary, he then explained the motives of the people of Kastellorizo and the ship owners. Apart from repeating that they wanted to avoid the payment of the compensation as a tax, he also explained that they planned to escape the island in order to avoid paying the tax.[41] In response, the Justice Ministry requested further information, acknowledging their ignorance of what was going on the island.[42]

Further complications arose when Cavit Bey, the officer charged with collecting the tax was unable to reach the island. This was because the Hayrettin steamship that was granted a special permit to transport him was immobilized in Yumurtalık due to lack of coal. In fact, coal was scarce throughout the empire at the time,[43] while a cholera outbreak in various locations required any available ship to go there.[44]

The inability of the Hayreddin to reach Kastellorizo added another layer of complexity to the affair, pertaining to the ability of the Ottoman state to exercise control over its coastlines and maritime domains. Steamships were few to begin with during the second half of the nineteenth century, and the coal shortage made things even worse.[45] As the government purchased large quantities of coal in the 1870s during the Ottoman-Russian war, ninety percent of the mining operators were unpaid.[46] The 1875 Ottoman bankruptcy meant that local suppliers could not be paid by the Ministry of the Navy, and coal was imported from Britain. This created the irony foreign steamships were powered by Ottoman coal from Ereğli (because the French boycotted English coal), and Ottoman ones with English coal.[47] These complications notwithstanding, it was eventually possible to secure enough coal for the steamboat to leave Yumurtalık near Adana and go to Kastellorizo.[48]

The amount of compensation, on the other hand, was determined after hard bargaining between the Ottoman government and the French embassy, reaching a compromise of 300,000 kuruş.[49] The people of Kastellorizo immediately informed Istanbul that this amount was beyond their ability given the poor conditions they faced. Emphasizing how the blockade had exacerbated their situation, they requested the payment to be made in three installments—again, an almost standard procedure for a province claiming impoverishment while obliged to pay an amount of money to the center. Their ships were seized as security for the eventual payment of the tax.[50]

Once collecting the first installment of the amount, Akif Bey informed Istanbul that he had no information on how to proceed with the payment. This miscommunication between the French and the Ottomans, meant that there was neither a payment procedure set, nor any French request. Fearing possible demands for interest incurred due to delays because of this, Akif Bey highlighted this issue. He went even further drawing attention to the question of how the amount would be converted from kuruş to francs. This would necessitate a request submitted by a financier (*sarraf*) who would then convert the money, but once again no such request was made. Following Akid Bey's report, Istanbul informed the French Ambassador that he needed to take action for the payment to be made promptly.[51]

The crew of the Hayrettin steamship were increasingly feeling apprehensive about their presence on Kastellorizo. They had faced various hardships and feared the reactions of the local since they were the ones enforcing the blockade by patrolling the island's maritime environs. They thus requested their dispatch back to their base, arguing to the Ministry of the Navy that their presence was no longer necessary. However, the Province of the Archipelago intervened, and informed the Ministry of the appearance of some bandits in Chalki (Herkit). The ship was thus ordered to remain in Kastellorizo and also deal with the situation in Chalki.[52]

After all three installments of the compensation money were collected, a further financial burden for the islanders emerged. Since the plundered wheat constituted import into the island, the Ottoman state demanded the customs duty.[53] After part of this new exaction was collection, by 1907 and after several petitions requesting mercy due to the dire economic conditions on the island, it was decided to absolve the people of Kastellorizo of the remaining amount of about 20,000 kuruş, effectively closing the affair after fifteen years.[54]

What is interesting here is the interplay between modern and premodern notions of justice and practices. On the one hand, we have the persistence on the strict application of the law which is observable throughout this case—very much along the lines of *Rechtsstaat* (law-based state). At the very tail end of the case, however, this modern mentality on the rule of law was counterbalanced by a very typically Ottoman, and one may say pre- or early modern, practice of showing sensibility to the pleas of its tax-payers requesting relief from their fiscal obligations.

Delving into the details of this affair opens a window into the spatial imagination of the Ottoman bureaucracy with reference to Kastellorizo. The island had traditionally been treated in many ways as an extension of the Anatolian mainland: it was organically tied to it in terms of provisioning, defense and security, or other means of state control. In this instance, however, we have a snapshot of a moment this came to halt. The Ottoman encounter with modernity, the forces of global trade (particu-larly so at a time global wheat markets were integrating), and insurance companies able to exert political pressure, created the conditions for a radical, if ephemeral, shift in how the Ottomans had viewed Kastellorizo. Istanbul completely reversed its hitherto practices and enforced a blockade of Kastellorizo thereby isolating it from the outside world. In this sense, the waters that connected Kastellorizo with the outside world, and the

Anatolian mainland in particular, became a hard border putting a stop to communication and transport.⁵⁵

On another level, and in addition to the importance of space, human agency is also part of the picture, particularly when nature and geography challenges collective or individual actors. Hence, the people of Kastellorizo did not consider plundering the ship a risk, and neither did they imagine that this affair would blow to such proportions. One may argue that this was a perhaps consequence of their limited experience and opportunities as islanders. This case is also important in examining the shifting dynamics of center-province relations, and how multiple bureaucratic agencies were involved in trying to deal with the affair. The collective punishment of isolation through blockade, turned insularity into a predicament for the residents of Kastellorizo: a prison within which, regardless of whether one was guilty of innocent of the accusation of plundering, all residents were forced to remain on account of their internal code of silence and communal solidarity against an often distant and absent state. While one may suppose that such bonds are more pronounced in the minimal insular space of Kastellorizo, these are by no means the prerogative of island societies.

Notes

1. Başkakanlık Osmanlı Arşivi (B.O.A.) İ.ŞD.118/7098, April 8, 1892. All archival references are from B.O.A..
2. Feridun Emecen, "Ege Adalarının İdari Yapısı," in *Ege Adaları'nın İdari, Mali ve Sosyal Yapısı*, İ. Bostan, ed., (Ankara: Stratejik Araştırma ve Etüdler Milli Komitesi Yay., 2001), 13; Yasemin Demircan, "Kikladlardaki Osmanlı İdaresini Şekillendiren Unsurlar ve 1670 Tahriri,*" Studies of The Ottoman Domain* 5:8 (Feb. 2015): accessed January 15, 2016, doi: 2147-5210. 71, 72.
3. Kolovos says that after their defeat at Lepanto in 1572, the Ottomans needed to strenghten their control over the Cyclades within the new historical conditions developed. Elias Kolovos, "Insularity and Island Society in Ottoman Context: the Case of the Aegean Island of Andros (Sixteenth to Eighteenth Centuries)", *Turcica* 39 (2007): 53.
4. İlhan Şahin, "Osmanlı Klasik Döneminde Ege Adalarında Nüfus ve Hareketleri," in *Ege Adalarının İdari, Mali ve Sosyal Yapısı*, İ. Bostan ed., (Ankara: Stratejik Araştırma ve Etüdler Milli Komitesi Yay, 2001), 136.

Also Herke, İleki, Kerpe and Anafi islands were subject to Rhodos as townships. *Ege Adalarının Egemenlik Devri Tarihçesi*, ed. Cevdet Küçük (Ankara; Stratejik Araştırma ve Etüdler Milli Komitesi, 2001), 19. A.Fuad Örenç, *Yakın Dönem Tarihimizde Rodos ve Oniki Ada* (İstanbul: Doğu Kütüphanesi, 2006), f. 52, 58.

5. For instance, it was tough to obtain control of political power due to different settlements, geographical structure, the geopolitical importance of the island and some other difficulties in broader islands such as Crete and Cyprus. See: Antonis Anastasopoulos, "Centre-Periphery Relations: Crete in the Eighteenth Century," *The Province Strikes Back İmperial Dynamics in the Eastern Mediterranean*, eds., B. Forsen and G. Salmeri (Helsinki: The Finnish Institute at Athens, 2008), 124. Antonis Hadjikyriacou, "Society and Economy on an Ottoman İsland: Cyprus in the Eighteenth Century" (PhD. diss., SOAS, University of London, 2011), 15.

6. Emecen, ibid, 25; C.AS. 914/39460, 1195 (1781); And some islands were conquored due to their particular products, for example: Midilli for its horses; Limnos for its plague curative soil, the island of Kiosk for its mastic, Simi for its sponges, Mamara for its marble. Nicolas Vatin and Gilles Veinstein, "Introduction" in *Insularités Ottomanes*, eds. Nicolas Vatin and Gilles Veinstein, (Istanbul: Institut français d'études anatoliennes, 2004), 7.

7. The Rumi population, which even today is limited in number, continue their lives within the opportunities provided by Greece to which they are mainly connected. Mehmet Bastıyalı, *Rodos ve Oniki Adalar* (İzmir: Arkadaş Press, 1999), 196.

8. Güner Doğan, *İngiliz ve Fransız Seyyahlara Göre 17. Ve 18. Yüzyıllarda Ege Adaları (Midilli, Sakız, Sisam, Rodos) ve Çevresi*, (Master's diss., Ankara Üniversitesi, 2008), 49; Evangelia Balta, "Ege'nin Ada Dünyası (15-19 yy)" in *Akdeniz Dünyası, Düşünce, Tarih Görünümü*, eds. Erol Özveren et al., (İstanbul: İletişim, 2006), 89.

9. HAT, 665/32324; Cf., the governor of Rodos Şevki Bey's petition, HAT, 665/32324-A.

10. 120 people were settled from the town of Katırlı in Kaş. MAD, 8457, Emecen, op. cit, 54. Thomas Dallam, *Early Voyages and Travels in the Levant, The Diary of Master Thomas Dallam (1599-1600)*, in J. Theodore Bent ed. (London: 1893), 33.

11. Fellows stated that the port of the island was full of large and small boats and that most of the residents of the island were both active and enterprising. Charles Fellows, *An Account of Discoveries in Lycia* 2 (London: 1840), 188–189.

12. Piri Reis, *Kitab-ı Bahriye*, in Bülent Arı, ed. (Ankara: T.C Başbakanlık Denizcilik Müsteşarlığı, 2002), 583. The island is mentioned in ancient literary sources as well due toits port. Adnan Pekman, trans., *Strabon*,

Geographika (Antik Anadolu Coğrafyası), Kitap (XII-XIII-XIV), (İstanbul: Arkeoloji ve Sanat Press, 2005), 247.

13. Fellows, ibid, 189; Piri Reis, on the other hand, records the existence of salt mining. Piri Reis, ibid, 583; fishing and sponge fishing are also recorded. Cuinet, ibid, 42.
14. İ.ŞD.118/7098, May 12, 1892.
15. İ.ŞD. 118/7098, 22 Ramazan 309 (20 April 1892).
16. İ.ŞD. 118/7098, 22 Ramazan 309 (20 April 1892).
17. İ.ŞD.118/7098, April 8, 1892.
18. Fatih Kahya, *Osmanlı Devleti'nde Sigortacılığın Ortaya Çıkışı ve Gelişimi*, (Master's diss. Marmara Üniversitesi, 2007), 36, 60-61. Ş. Özdemir Gümüş, "Osmanlı Deniz Ticaretinde Hukuki Bir sorun: Kaza Yapan Geminin Malları," *Folklor/Edebiyat* 3 (2012): 115; Fatih Kahya, "Osmanlı Devleti'nde Deniz sigortacılığı", in *İskeleye Yanaşan Denizler, Gemiler ve Denizciler*, Orhan Berent et all. (İstanbul; İletişim Press, 2013), 341. Murat Baskıcı, "Osmanlı Anadolusunda Sigorta Piyasası: 1860-1918," *A.Ü., Siyasal Bilgiler Fakültesi Dergisi* 57: 4 (2002): 3. Hatime Kamil Çelebi, "Osmanlı Devleti'nde Sigortacılığa Yönelik Dini Yaklaşım," *Ekonomi Bilimler Dergisi* 4:1 (2012): 107, accessed January 8, 2016, doi: 1309-8020.
19. İ.ŞD.118/7098, April 8, 1892.
20. İ.ŞD. 118/7098, 22 Ramazan 309 (20 April 1892).
21. Fatma Şimşek, "Anadolu Sancaklarında Mütesellimlik Kurumu (XVIII. Yüzyıl)," (PhD diss., Akdeniz Üniversitesi, 2010), 80–81.
22. İ.ŞD.118/7098, Fi 14 Şevval 309 (12 May 1892; İ.ŞD.118-7098, 21 Şevval 309 (19 May 1892); BEO, 12/837, 3 Zilkade 1309 (30 May 1892).
23. DH.MKT. 1961/33, 18 Zilkade 1308 (23 June 1891).
24. BEO, 179/13403, 22 Mart 1309 (3 April 1893); BEO, 183-13661, 10 Nisan 1309, (22 April 1893); DH.MKT. 1988/42, 22 Muharrem 1310 (16 August 1892).
25. BEO, 198/147772, 4 Şevval 1310 (11 May 1893).
26. BEO, 188/14056, 8 Şevval 1310 (25 April 1893), DH. MKT., 2001/87, 22 Safer 1310 (15 September 1892).
27. BEO, 65/4871, 28 Eylül 1310 (10 October 1893); BEO, 191/14264, 14 Rebiülahir 1310 (5 November 1892); DH.MKT. 1964/99, 27 Zilkade 1309 (23 June 1892).
28. BEO, 188/14056, 23 şevval 1310 (10 May 1893).
29. BEO, 191/14264, 14 Rebilülahir 310 (5 Octaber 1892).
30. BEO, 188/14056, 8 şevval 1310 (25 April 1893), BEO, 11/755, 29 Mart 1209 (10 April 1893), BEO, 11/755, 12 Nisan 1309 (10 April 1893), BEO, 185/13837, 3 Nisan 1309 (15 April 1893).

31. BEO, 191/14264, 14 Rebiülahir 1310 (5 November 1892), BEO, 188/14056, 8 şevval 1310 (25 April 1893), BEO, 185/13837, 5 Nisan 1309 (17 April 1893).
32. BEO, 11/755, 28 Şevval 1209 (26 May 1892), BEO, 183/13663, 25 Ramazan 1310 (12 April 1893).
33. Y.MTV, 75/208, 3 Mart 1309 (15 March 1893); BEO, 11/755, 28 Şevval 1309 (26 May 1892); BEO, 11/755, 9 Mart 1309 (21 March 1893). There were several complaints about the working methods of the commission. For example, that the amount of compensation was established without the commission arriving to the island. BEO, 65/4870, 18 Mart 1309 (30 March 1893).
34. Y.MTV, 75/208, 3 Mart 1309 (15 March 1893).
35. BEO, 197/14742, 23 Şevval 1310 (10 May 1893).
36. For the role of petitions early modern Ottoman governance and legitimacy see Boğaç A. Ergene, "On Ottoman Justice: Interpretations in Conflict (1600-1800)," *Islamic law and society* 8 (2001), 52–87.
37. BEO, 11/755, 9 Mart 1309 (21 March 1893); Cuinet, ibid, 42.
38. BEO, 11/755, 28 Şevval 1309 (26 May 1892).
39. ŞD, 2616/5, 8 Şevval 1310 (25 April 1893), BEO, 11/755, 12 Nisan 1309 (24 April 1893), BEO, 205/15330, 2 Zilkade 1310 (18 May 1893).
40. BEO, 11/755, 12 Nisan 1309 (24 April 1893).
41. BEO, 11/755, 10 Mart 1309 (22 March 1893).
42. BEO, 205/15330, 2 Zilkade 1310 (18 May 1893).
43. BEO, 179/13360; 21 Mart 1309 (2 April 1893), BEO, 11/755, 28 Şevval 1309 (26 May 1892).
44. BEO, 11/755, 28 Şevval 1309 (26 May 1892), BEO, 182/13596, 29 Mart 1309 (10 April 1893).
45. A. İhsan Gencer, *Türk Denizcilik Tarihi Araştırmaları* (İstanbul: Kıral Matbaası, 1986), ff 19, 32.
46. Donald Quatert, "State Discipline and Villigers' Resistance to Mine Work in the Zonguldak Coalfield (1820-1920)", in *Popular Protest and Political Participation in the Ottoman Empire*, eds. Eleni Gara et al., (İstanbul: İstanbul Bilgili Üniversitesi Press, 2011), 295–296; Donald Quatert, *Osmanlı İmparatorluğu'nda Madenciler ve Devlet,* trans., Nilay Özok Gündoğan, Azat Zana Gündoğan, (İstanbul: Boğaziçi Üniversitesi Press, 2009), 347–350.
47. İlhan Ekinci, "XIX. Yüzyılda Osmanlı Deniz Ticaretinde Değişim ve Tepkiler," *Tarih İncelemeleri Dergisi* 21: 2, (2006): 46, 51. Donald Quatert, *Osmanlı Devleti'nde Avrupa İktisadi Yayılımı ve Direniş (1881-1908)*, trans. Sabri Tekay (İstanbul: Yurt Yayınları, 1987), 51–52.

48. BEO, 198/14777, 24 Şevval 1310 (11 May 1893).
49. BEO, 241/18048, 28 Zilhicce 1310 (13 Temmuz 1893).
50. BEO, 11/755, 28 Şevval 1209 (26 May 1892), BEO, 594/44506, 8 Şevval 1312 (4 Nisan 1895).
51. BEO, 264/19770, 12 Safer 1311 (25 Ağustos 1893).
52. BEO, 210/15735, 17 Mayıs 1309 (27 Mayıs 1893).
53. BEO, 2838/212839, 4 Rebiülevvel 1324 (28 April 1906).
54. BEO, 321/22574, 14 Teşrin-i evvel 1309 (26 October 1893), DH.MKT, 866/35, 24 Rebiülevvel 1322 (8 July 1904); İ.RSM, 24/17, 3 Rebiülahir 1324 (27 May 1906).
55. Guy Mercier, "Etude de L'insularité (Rapport sur le Premier Theme)," *Norois* 37:145 (P1990), 11.

The Ottoman Peloponnese before the Greek Revolution: "A Republic of Ayan, Hakim, and Kocabaşı" in "the Sea of Humans and Valley of Castles"

KAHRAMAN ŞAKUL

Introduction

The Peloponnese is presently an island because of the Corinth Canal completed in 1893. If, in this case, a 6.4 kilometer-long canal is what distinguishes a peninsula from an island, could there be a mental geography that classifies the Peloponnese as the latter, rather than the former? The seventeenth-century traveler Evliya Çelebi would confidently confirm this suggestion. He narrates how the Venetians had cut a canal at Corinth to turn the province into "an island" (*ada*) as a measure against the looming Ottoman threat. However, the Spanish king all of a sudden conquered this "safe island" (*cezire-i asude-hal*). He filled up the canal so that it was once again connected to the mainland, rendering the Ottoman conquest of the "island" all the easier. With a canal or not, the Morea was always an island in Evliya's mind.[1] Having this idea as its point of departure, this article ventures to explain what it meant for the Ottoman Morea to be perceived as an island.

The Ottoman Turkish language does not distinguish between islands and peninsulas. The most commonly employed word for island, *cezire*, is an Arabic loan word that implies not only the surrounding of land by sea, but any kind of physical/geographical isolation. In Braudelian terms peninsulas, oases, and mountains are "almost islands" that the sea does not surround.² In the relevant entry of the Arabic-Turkish dictionary he translated, Asım Efendi pointed out *el-Ceziretu'l-Hadra* took its name from the unique and graceful qualities that distinguished it from its immediate surrounding in Andalusia.³ The fifteenth-century chronicler Neşri narrated that the frontier lord Turhan Bey had described to Murad II the Morea as an island (*ada*) cut off by the Hexamilion and "completely surrounded by the sea."⁴ Evliya described the castle of Chlemoutsi as an island-like place (*cezire yer*) isolated from the rest on the top of a mountain.⁵ This linguistic ambiguity in Ottoman Turkish, then, denotes a metaphorical definition of island, which is not much different from the definition proposed by cultural anthropology: an island is "not just an area surrounded by the sea, but a social, political economic, cultural unit with its own character and development, an integral part of a greater unit which may include other islands and mainlands."⁶

Often labeled as miniature or matchbox continents, these insular spaces imply both isolation and connectivity.⁷ Antonis Hadjikyriacou has proposed three distinct but overlapping settings to study Ottoman islands: the Mediterranean, the Ottoman and the local.⁸ These contexts may have a multiplicity of fiscal or administrative systems, and may include seemingly exclusive or opposing categories (ethnic, linguistic, religious), depending on the circumstances. Combining this approach with Geertz' concept of "thick description"⁹ offers an understanding of insularity that has the potential of unraveling the functioning of a network consisting of fiscal-administrative, economic, cultural, and ideological systems. Needless to say, the idea of "thick description" is used here as an analytical tool for understanding the issues of isolation and connectedness in the multi-layered insularity of an Ottoman island.

It is my contention that the Morea was a "perceived island",¹⁰ and treating it so allows us to better understand the general trends in administrative and fiscal developments in the Ottoman Empire, as well as ideological shifts in the Mediterranean world during the eighteenth century. While the Peloponnesian insularity was not an integral and necessary condition for the outbreak of the Greek Revolution, political crises and ideological shifts still loomed large in the unfolding of events that culminated to this outcome, thereby effectively ending the administrative and political distinctiveness of the Morea.

The Mediterranean Context

The Mediterranean Sea was "a watery continent"[11] with many countries including the Adriatic and Ionian as well as the Morea that stood in between the Ottoman world and the Adriatic. The interaction within the Mediterranean basin was as much significant as the relations with Northwestern Europe in the 1820s revolts. Pushkin, for instance, showed no hesitation to view "the four horsemen" (Rafael del Riego, Guglielmo Pepe, Muraviev-Apostol, and Alexandros Ypsilantis) as the embodiment of the revolutionary spirit that aimed "to bring political freedom and a constitution for Spain, Naples, and Russia and national independence for the Greeks," respectively.[12] Neapolian and Piedmontese revolutionaries fought in the uprisings in the three afore-mentioned peninsulas, thereby forming the living link between them, while the Spanish Constitution of 1812 provided the template for the long-esteemed constitutional regime.[13] The mechanical similarities embedded in this movement raging through Spain, Naples, and the Morea included "the impact of the Napoleonic wars, assorted grievances, secret society preparation, military initiators, the *pronunciamento*, liberal aspirations, heroic postures, inner conflicts, cultural and religious adaptations."[14] Thus, a Mediterranean perspective is necessary to explain the reconfiguration of the Peloponnese as a political space in the aftermath of the promulgation of Greece.

Morea's insularity within the Mediterranean context was based on its political subordination to the Ottoman rule and commercial integration to a "world-system" in the Ottoman-Venetian frontier. But this was soon to change after the French Revolution. As the Adriatic turned into a new phase as a frontier zone after 1797, the local magnates had to maneuver in the dangerous waters of Mediterranean politics to make the best of the situation; this produced a number of internationally prominent figures in this region. Ali Pasha of Yanya (Ionnina) was the most obvious case of political expediency in this frontier region. Double-dealings with foreign powers, dubious allegiance to the Sublime Porte, and increased political influence and commercial endeavors along the Dalmatian coasts defined his thwarted enterprise.[15] Among many "trans-imperial" actors in this frontier was the *kocabaşı* Panagiotis Benakis from Kalamata, one of the leading figures of the Orlov Revolt of 1770 in the Morea. His son

Liberakis Benakis took refuge in Corfu and entered in Russian service. He became the Russian consul-general in Corfu, owing to his lineage and the support of Vasiliy Stepanovich Tomara (1746–1819), the Russian ambassador to the Porte of Geek parentage. As a former Ottoman Moreote, this Russian consul of Corfu was very conversant in local politics and imperial rivalries. He cultivated good relations with Mustafa Reşid Efendi, the Ottoman Superintendent of Corfu, to whom he once lent money.[16]

In this context of the Morea's immediate environs, Count Antonio Maria Capodistria, father of Ioannes Capodistrias who would later become the first president of Greece, was another local-international actor who tried to preserve the traditional political and social system in the Ionian Islands in the midst of intense imperial rivalry. He was a vehement supporter of the much-discussed and disputed constitution of the Septinsular Republic that favored the rule of nobility under Ottoman-Russian protection; he joined the pro-Ottoman Ionian delegation sent to Istanbul in 1802.[17] In the backdrop of the internationalization of the entire region, the political dimensions were affected as well.

The Revolutionary wars and the suppression of Venice in 1797 eliminated the French and Venetian shipping from the Mediterranean, and benefited Greek maritime trade.[18] Greek-owned ships sailing under Ottoman flag met the demand for neutral shipping, and constituted 80 percent of the ships heading to ports in the Western Mediterranean in 1780–1810.[19] Harlaftis described the expansion of Ottoman-Greek maritime trade as the "Eastern grain invasion" since most of these ships supplied Spain with grains particularly from the Black Sea region. Greek captains from *Çamlıca* (Hydra) went as far as the Caribbean with mixed crews of Turks/Muslims and Greeks.[20] The Morea took its share in the Mediterranean trade as the exporter of raisins, olives, olive oil, wool, and grain in exchange for mainly English and French manufactured goods, as well as coffee, tea, indigo, and cochineal through the ports of Balyebadra (Patras) and Vostiçe (Vostitsa).[21]

The Revolutionary and Napoleonic wars also had political and social consequences in the province. The French invasion of the Ionian Islands and the coastal strip of Dalmatia meant the opening of the Adriatic frontier. The Ottomans struggled to fill the vacuum in the Adriatic and

Dalmatia after the demise of Venice, culminating in alliance with Russia and Britain against France. This would result in the creation of the Septinsular Republic under Ottoman-Russian protection in 1802.[22] The unruly lords of the mountains of Epirus and the Dalmatian coasts (Tsamides, Souliotes, and Khimariots, among others) were a source of trouble that was typical issue of a frontier region such as the Morea.[23] Prior to the French invasion of Egypt, the Morea was one of the Ottoman provinces where the refugees from the Ionian Islands under French occupation preferred to reside. The Sublime Porte recognized them at first as French subjects, and allowed only those with a Moreot wife to stay in the province as long as they had a guarantor (*kefil*). After the declaration of war on France, though, the Porte allowed the Ionian refugees to become Ottoman subjects and recruited volunteers for the Ottoman fleet operating in the Adriatic.[24]

The Moreots were accustomed to propagandist pamphlets and proclamations since the 1790s as part of the same process. Ottoman authorities consternated over Rhigas constitution, seditious letters of Napoleon addressing the Maniotes as well as inflammatory French pamphlets. The governor of the Morea, Hasan Paşa, alerted Istanbul about the subversive French activities. He, however, misunderstood a French pamphlet about the impending Egyptian expedition as a clear sign of French designs on the Morea.[25]

The local magnates and priests were frightened by the duplicity of the French, whom had showed no respect for the neutrality of Venice. Apprehended by the French annexation of Venice, the communal representatives feared that the "Jacobin bandits" would invade the province in complete disregard of the Ottoman neutrality as well. Thus, they asked of the governor to demand 5,000 troops from Istanbul in order to avoid the fate of Venice.[26] By the autumn of 1798, the Ottoman-Russian joint fleet arrived in the Adriatic to fight the French in the Ionian Islands. The central place of the Morea within the regional network is exemplified by the correspondence of Lalalı (the Lalliote) Mustafa and Seydi Ağa in which they contended that they could have the Zantiotes detain the French garrison. They were counting on their good relations with Zante based on grain trade since the Morea traditionally supplied the Ionians with grains.[27] As a matter of fact, the French garrison in Corfu too had to buy provisions from Yanya and the Morea in 1797 by the permission of the Porte.[28]

The Adriatic frontier became an arena for imperial rivalry between the Ottomans, French, English, and the Russians at the turn of the eighteenth century. The Ottomans tried to involve Britain in Ionian affairs particularly during the rebellion in Corfu in 1803 in order to offset the increasing Russian influence over the Greeks. Besides the propagandist pamphlets and proclamations, the revolutionary vocabulary was also gaining popularity in this frontiers zone. The Corfiote rebels, for instance, were "Jacobins" in the eyes of the Ottomans, British, and the Russians. Under such precarious conditions, the Ottomans thought it feasible to deploy British troops in the Morea against the French threat once the rebellion in Corfu was over. This is striking when one considers the Porte's outright rejection of foreign intervention to negotiate with the Greek revolutionaries during the Greek Revolt.[29]

One intriguing case of elite mobilization across confessional divides warns us against the danger of overemphasizing the linguistic and ethnic divides in the Morea. In 1808, Muslim and non-Muslim notables of the province conspired to make the Morea a French protectorate. While this move was in contrast to their anti-French stance almost a decade ago examined above, it is indicative of the shifting allegiances and alliances that such agents had during a period of transition. Threatened by the tightening grip of Ali Pasha of Yanya whose son Veli served as the governor of the Morea in 1807–12, there was indeed nowhere for Peloponnesian elites to turn to except France. Istanbul had to deal with two coups, a civil war and two murdered sultans (Selim III, Mustafa IV) in just two years (1807–08). Still worse, it was in the midst of yet another war with Russia. France, by contrast, bitterly defeated Prussia, Austria and Russia in recent wars. The plan proposed by the local notables who signed as *Confédérés de Morée* envisioned a Peloponnesian confederation. Each district would have an administrative body of three notables of Muslim and Christian origins. Remarkably, the notables labeled the Morea as their *patrie*, which indicates a local identity among the magnates that transcended religious differences and was based on territory. The plan was null and void after the British conquered Corfu in 1809 and the Porte removed Veli Pasha in 1812. It, nevertheless, implies that the magnates preferred to keep the internal system intact and that they were ready to shift allegiance from one empire to another.[30] Apparently, the protectorate system established in the Ionian Islands in 1802 served as the model for the local notables.

The Ottoman Context

The Morea was a distinctive province in several ways despite the application of usual Ottoman administrative practices. Although the province was in a frontier region, many of its resources were earmarked for the Ottoman dynasty in the form of tax-farms, which further complicated the administrative setting. Sultan Selim III, for instance, had to create two offices concerning this frontier region with special duties: the overseers of the Morea and Corfu (*Mora Nazırı*).[31] Furthermore, its commercial links with Istanbul and the ports of the Black Sea and the Mediterranean were through maritime routes. By the same token, the military conflicts over the control of the province and its vicinity always involved naval operations.

After the Ottoman conquest, the agricultural lands of the province had at large been given to the prebendal *timar* system to support the provincial cavalry. Since the 1650s, it joined the 22 islands and coastal territories under the jurisdiction of the Ottoman grand admiral who paid an annual lump-sum (*salyaneli*) as tax.[32] Two major wars with Venice (1645–69 and 1685–99) undermined the economic wellbeing of the province because of the population decline and destruction of property. After the second conquest in 1715, the Sublime Porte attempted to impose central rule by reintroducing the *timar* and inviting former Muslim and Turkish residents back to take up their former lands. The introduction of tax-farming in the Morea in 1747 created the fiscal-administrative framework that was present there on the eve of the Greek Revolt.[33]

The new political, fiscal and geographical configuration of the empire was based on an amalgamation of local households, known as *ayan* dynasties.[34] In the words of Ali Yaycıoğlu, a process of localization, privatization, and communalization institutionalized the role of *ayan*s as intermediaries between state and society. This is best described as the transformation of "*ayan*-hood as natural leadership" into "*ayan*-ship as a formal office."[35] Such local dynasties were political, military, and economic enterprises, with the overlapping functions of provincial rulers, trading companies, or landlords with associates in Istanbul.[36] The Sublime Porte and the local magnates ceased to represent "the center" and the "periphery", respectively in this flexible and integrative polity based on contractual relations.[37]

The case of Ebubekir Beg, a notable from Gördos (Corinth), exemplifies the multitasking nature of the local foci of power and calls into question the classical center-periphery paradigm. He was to supply provisions to the Russo-Ottoman joint fleet in the Adriatic, the garrison of the castle of Anaboli (Nauplio), and the new troops dispatched to the Morea against a possible French attack. In addition, he had to supervise the reinforcement of a number of castles in the province. The Porte desperately needed his services so his repeated requests to be released from the commissariat services fell on deaf ears.[38] Ebubekir Efendi, along with Azmi Efendi, the Overseer of the Morea, even participated in a meeting in Istanbul in order to face the accusations of the Russian ambassador concerning the low-quality of the hardtack biscuit Ebubekir had locally delivered to the Russian navy.[39]

Various districts in the Morea were assigned to high officials or members of the Ottoman dynasty. These districts were free (*serbestiyet*) from the intervention of the Governor, and exempt from some of the extraordinary taxes. Absenteeism was the norm among the owners of these sinews as well as the Istanbul-based shareholders (*esham*-owners) of those lucrative tax-farms of the tithes, the sheep tax, and the olive oil. The last cadastral survey of the Morea conducted in the aftermath of the re-conquest in 1715 formed the basis of taxation. Therefore, the *ayan* had a good portion of manipulative power in the reassessment of tax rates and distribution of taxation.[40] In view of their influence and local knowledge, the local elite—Muslim *voyvoda*s and Christian notables (*kocabaşıs*)—were the perfect sub-leasers.[41]

The governor with his seat at Tripoliçe (Tripolis) represented the political and military authority in the peninsula with two councils functioning under him. The Council of the Morea (*Mora Divanı*) was the advisory body of functionaries. By contrast, the Council of the *Ayan*s (*Mora Ayanları*) consisted of two Muslim and two non-Muslim representatives. The dragoman of the Morea (*Mora Tercümanı*) was the only non-Muslim member of the former council, which made him the chief mediator between the official authority and the Christian community.[42] A confidential letter of Governor Mustafa Pasha, most likely addressing the deputy grand-vizier in Istanbul, epitomizes most of the fiscal-administrative complexities of the province:

> You are pointing out that I have been granted larger authority [*ruhsat*] than my predecessors and that I should act with confidence in executing the office in complete freedom. Very well! But, the *cezire* of the Morea is composed of 20 *kaza*s. The owners of 16 of them are evidently known. All in all, only four *kaza*s belong to the *cezire*. It needs not mentioning that the course of the events in sixteen *kaza*s is not determined by the governor. The subject of the sublime edict that just arrived is the inspection of the accounts of the Hacı Hasan Ağa, the *voyvoda* of Karitena, concerning the tax-farm of the poll-tax and the accounts of Deli Yanni (Deliyannis), the *kocabaşı* of Karitena. If I set about to deal with it today in conjunction with the decree, they will produce a contradicting decree in a matter of less than ten days. Also, the lady [Beyhan Sultana] who owns the tax-farm will be annoyed. What is the solution to this my Sultan?[43]

When Mustafa Paşa reminded the Sublime Porte about his weak position vis-à-vis the *kaza*s, he meant that these districts simply fell outside of his jurisdiction as governor of the province. Therefore, he sought to obtain a "license", authorizing him to investigate the matter and forestall a possible reproach of Beyhan Sultana (1765–1824), the beloved sister of Selim III and the owner of the lifelong usufruct of the districts in southern Peloponnese, namely, Karytaina and Messenia. One of the wealthiest princesses in Ottoman history, Beyhan Sultana farmed these resources through her *voyvoda* Hacı Hasan Ağa whom she appointed in 1802 when the former *voyvoda* Hüseyin Ağa had proved to be too oppressive.[44] The Deliyannis clan of these districts also protected the regional interests of Beyhan Sultana. This nexus of fiscal and political power, linking a provincial with a member of the imperial family, highlights the extent to which a provincial class of notables could be integrated into Istanbul-centered politics.[45]

It is amusing to realize the extent of the effort spent by foreign travelers touring in the Morea in order to comprehend the administrative structure of the land.[46] The Porte was no less confused than the foreign observers. This confusion is best reflected in two Law books of the Morea (*kanunname*), dated 1793 and 1812. They draw a clear picture of the fiscal and administrative framework of the province, the role of the local elite, the Governor, and popular grievances. Both law books condemned the oppression of the governing pashas as well as the *voyvoda*s that compelled:

> the inhabitants of each *kaza* to seek ways to obtain a decree granting free status as that of Patras so as to find a safe haven. And even those *kaza*s under the supervision of royal princesses are striving to obtain decrees granting them free-status and exemptions. As a result, only 5–6 *kaza*s remained to shoulder the oppressive exactions of the Governors, while other *kaza*s were almost detached from the Morea. Meanwhile, the subjects of the tax-farms of the royal princesses, privileged by the granting of free-status, are suffering from the oppression of their appointed *voyvoda*s. At present, one part of the Morea is devastated by the oppression of the Governors, and the remaining part is ruined by the exactions of the *voyvoda*s.[47]

These law codes attempted to end corruption on the local level by a series of measures. Regardless of the fiscal status of a unit, the Governor would have all of the *kaza*s. A new cadastral survey was meant to end the injustice in taxation. The 1812 law code followed the former almost in verbatim, suggesting that it had become something of a bureaucratic tradition to issue such laws after each war.[48]

The rhetorical aspect of the laws on the Morea leaves the impression that the Porte generously acknowledged the shortcomings of its politico-fiscal and administrative system, but could do nothing beyond warning the local functionaries against corruption and oppression. Naturally, the issues of transgression and venality were not limited to the province of the Morea. However, the scale of such problems seems to have been larger in this island by the reactivation of the Adriatic frontier after 1797. It was hard pressed by the unfolding imperial rivalry in the form of continuous naval operations. Low official prices and cumbersome payment methods in the military logistics system caused many complaints among the local peasants as well as the impoverishment of the "*ayan* dynasties" (*hanedan*), at least from the viewpoint of the governor of the province.[49]

The island's connectedness to the outer world through legal and illicit trade must have alleviated the burden of taxation in a time of agricultural boom. Enormous smuggling notwithstanding, even the official data concerning the extortionate quotas of provisions the Morea sent to the Russo-Ottoman joint fleet hint at significant production levels of commercial produce.[50] A study of Anavarin (Navarino) and its vicinity suggests that in the course of the eighteenth century, Greek peasants substantially increased the amount of land under their control, and that rural tax-farms were likely to have increased in most of the province.[51]

In short, the new system did not necessarily lead to political decentralization, but caused peasant indebtedness, widespread rebellion and banditry. The *ayan*s usually prepaid taxes only to charge interest on them and collected various service fees; hence, the saying "the country labours under three curses: the priests, the kocabasis, and the Turk."[52] In his treatise on the so-called "Orlov" Revolt in 1770, Süleyman Penah Efendi (born in Tripoliçe) condemned the practice of assigning fiscal-administrative duties to locals as a major reason for the rebellion and the ensuing Greek emigration. He shared the opinion of his Christian compatriots, when he perceptively described such provinces as the Morea as "republics of *ayan*, *hakim* [administrative, legal, and religious authorities], and *kocabaşı*, who considered the Ottoman provinces as their family inheritance."[53]

The Local Context

Evliya Çelebi was a keen observer of several features that made the Morea a miniature continent. These features ranged from the variegated fauna, flora, and geography to language and cultural variations. The

Rufiya (Alfios) river that flowed into the Mediterranean near Fenari particularly stunned him, as he could not explain the presence of such a large river in an island in the middle of the sea.[54] After his grand tour in the island, he concluded, "in short, the island of the Morea is a sea of humans and a valley of castles on this earth."[55] Language was an obvious divide in this miniature continent and Evliya Çelebi pointed out that the non-Muslim subjects (*re'aya*) of the island were all Greeks and Albanians. He divided the locals into five groups along ethno-linguistic lines, contrasting the refined and articulate Greek spoken in Mezistre with the "unintelligible" dialect of the Maniotes. The Lakonians of Nauplia and Benefşe (Monemvasia) were also distinguished by their strange dialect as much as their tough bodies. Albanians concentrated in the north, whereas all Muslim subjects spoke Greek.[56] Jews, Franks, black (*kara Arab*s, *zenci Arab*s), and white (*Frenk*) slaves enter occasionally in his textual map of the Morea. A revealing proof of the island's connectivity is the Algerian-style short pants and slim shirts of the young men, Christian and Muslim alike.[57]

Ethnicity, religion, and language were interrelated social markers in the minds of the Ottoman Turkish elite. For instance, Albanians were defined by their ethnic name even if they were Muslims.[58] Evliya observed in the 1670s that the Turkish Muslim military administrative class usually preferred to live in the fortress towns in relative isolation from the Greek majority of the rural population.[59] Thus, Greek was considered to be the language of a subject people. William Gell, an English traveler, remarked that the etiquette had required the governor of the Morea to speak in Turkish in public, although he preferred Greek in his secret informal meeting with himself.[60] A book of etiquette from the early eighteenth century showed contempt for Muslims who spoke languages other than Turkish in Istanbul.[61] This insistence on using the language of the state could be taken as an indication of genuine conversion. All these are strongly reminiscent of the processes found in another "miniature continent", Crete.[62]

Süleyman Penah Efendi also drew attention to the linguistic (*lisan*) and ethnic (*cins* and *millet*) variety in the island. In his opinion Greek-speaking peasants were peaceful and obedient, whereas Albanian-speaking ones were ill mannered and rebellious, apparently due to their native tongue. According to him, Albanian was a rough language devoid of any grace. Penah Efendi argued that the nurturing of any given people was bound to the knowledge of the language of the state, and accordingly insisted that the Porte should make Albanians adopt Turkish. He justified this policy by the Spanish example. According to him, the Spanish married native women and raised bilingual sons who made Native Americans forget "the American language" in one generation.[63]

Penah's ambitious, if early, *mission civilizatrice* project involved the imposition of the Turkish language on Albanian clerics and community leaders. Russia was another source of inspiration. He maintained that Russia had recruited many Greek boys to train them in seditious activities against the Sublime Porte. Thus, the Porte should also educate a number of young Tosk Albanians from Delvina and Avlonya (Vlorë) in Istanbul. These new Ottomans would register all the boys of the Albanian nobility over 15 and send them in the pasha households throughout the empire for training. The Ottoman Empire would thus tame the unruly Albanian warriors and gain obedient troops.[64] Such proposals foreshadow Abdülhamid II's projects for the incorporation of tribal elites through education.[65]

Penah Efendi was obviously accentuating his resentment against the alienation of the traditional elite—Muslim and non-Muslim alike—in the Morea during *Alvanokratia* (the period of Albanian rule). By these measures he was hoping to pacify the Albanian mercenaries, who settled unlawfully in the province after suppressing the Orlov rebellion.[66] The Albanians caused much disturbance to social and economic life of the province, so much so that the Sublime Porte had to arm the Christian subjects to eject them from the Morea. It seems that the ferocity of the Albanian troops created an entrenched ethnic and religious polarization in the province in the long run.[67]

Some Christian notables of the Morea considered the tax-farm system as the end of "democracy" (i.e., local autonomy).[68] Many subjects resented the system, as revealed in popular rumors. A contemporary Greek chronicler recounted the current belief that Haseki Hacı Ali, the tyrannical *voyvoda* of Athens, was able to purchase the town at auction as a tax farm (*malikane*), allegedly thanks to a love affair with Esma Sultan (1726–88), the daughter of Ahmed III.[69] Conversely, most historians agree that the Christian notables in the Morea enjoyed a level of political autonomy much greater than power-holders in other provinces of the empire. This was due to local collaboration in the Ottoman re-conquest of the Morea in 1715. Commenting on the revolutionary processes of the 1820s in the Peloponnese, Stamatopoulos views the non-Muslim local elites and their increased political role as the decisive factors, rather than the ship-owners of the Aegean islands and the klephts of continental Greece.[70]

The first Ottoman permanent ambassadors sent to London and Paris in 1793 were from the Morea. Apparently, the choice of the Porte was not coincidental. The Moreot Muslims were considered more familiar with Western culture and languages. Yusuf Agah Efendi, ambassador in London (1794–97), was the son of the afore-mentioned Süleyman Penah Efendi. His chief secretary, Mahmud Raif Efendi, was from the Morea as well; he also served as the diplomatic representative in the Russo-Ottoman fleet in 1798–99. As a sign of the Porte's pro-British diplomacy, "*Ingiliz* Mahmud" was in charge of Ottoman foreign affairs in 1800–5. He was the first reformer to be murdered by rebel forces in 1807, ending the reform project known as *nizam-ı cedid* (the new order).[71] Seyyid Ali Efendi, the first permanent ambassador to Paris, was also born in the Morea and owed his appointment to the post to his knowledge of French. He was a prominent figure among the reformists who organized the plot that put Mahmud II on the throne in 1808.[72]

Ancient Athens was quintessential to philhellenism and European fascination with the Morea. Less known is the story of the Ottoman Muslim intellectuals who were captivated by antiquity. The achievements and monuments of antiquity never escaped the attention of Katip Çelebi (*Hacı Halife*), Hüseyin Hezarfen, Müneccimbaşı, and the renowned traveler Evliya Çelebi.[73] Mahmud Efendi, a native from Thebes or Euboia, wrote the history of Athens in 1738. Served many years as the jurisconsult of the city, he had an acute sense of the achievements of ancient Athens as well as its difficult relations with the Morea. In his *History of the City of Philosophers*, Athens stood out with its ancient buildings, wisdom, and refined culture, whereas the Morea was defined by tough terrain, unsafe roads, bandits, and troublesome rulers (*shah*s) for the most part.[74] Most significantly, Mahmud Efendi was keen on historical continuity. He understood the history of the peninsula as a seamless past, regardless its various conquerors. He remarked that just like Romans, Gog-Magogs, and Summerian Jews so were *Türk* and *Yunan* also descendants of Japheth (*Yafes*).⁷⁵ Furthermore, he "Ottomanized" the past experience through a liberal application of the Ottoman political-fiscal terms, such as *padişah, hakim* (ruler), *sultan, serasker* (commander), *bey, cizye* (polltax), *öşr* (tithe), *rüsumat* (dues) and so forth, in narrating antiquity.⁷⁶ In his understanding, then, there was historical and ethnic connectivity that tied the Morea to the Ottoman world; this is best observed in the last section of his book, where he narrated extensively the Ottoman reconquest of the Morea as an eyewitness.⁷⁷

Muslims of the Morea had come a long way, from the times of Mahmud Efendi to the turbulent 1820s, in appreciating ancient Greek civilization. It is revealing that the British traveler Gell left the *Athenian Society of Philomusae* in reaction to its refusal to extend membership to the Muslims of the city.⁷⁸ The Muslims were, thus, effectively excluded from the glorious past. By the same token, "the only example of engagement with the revolting Greeks' worldview" on the part of the Ottomans was the treatise submitted by a certain Asimaki (probably Asimakis Zaimis) from the Morea to the Sublime Porte in 1823. He discussed the causes of the revolt and proposed possible solutions, which Sultan Mahmud II cared to read and forward to the Imperial Council.⁷⁹

In the midst of scenes of savagery and carnage during the Greek Revolt, there were still moments of mutual support and help. Christian and Muslim friends strove to help one another on several occasions. A Maniote "bandit" warned his old acquaintance, sheikh Necib Efendi of Tripoliçe, that "the Greek nation" (*millet*) was all united and the Muslims should watch out. Prominent Muslims sent secret letters to the *kocabaşı*s of Kalavrata and Patras, telling them not to go to Trapoliçe to attend the meeting with the governor so as to escape a likely execution. On another occasion, Greek peasants refused to hand over a certain *timar*-holder to the insurgents, and escorted him back to the fortress of Anabolu. In a most moving case, the insurgents from the nearby villages gave up the idea of massacring the Muslims of Benefşe (Monemvasia), since many of their relatives were married to the Muslim men of the town.⁸⁰

Conclusion

Sheer violence radicalized the peninsulas of the Mediterranean basin—Italy, Spain, and the Balkans—after the French Revolution. The full effects of the internationalization of the Adriatic from 1797 onwards were evident in the confessional bloodshed of the 1820s. Greek insurgents had the necessary military capabilities and the political determination. Thousands of Greeks served in the armies of Russia as well as the French and British armies in the Ionian Islands after 1797. These disciplined troops contributed immensely to the Greek Revolution.[81] Initially, the Sublime Porte deemed the Greek Revolt in the Morea as an extension of the revolt of Ali Pasha of Yanya. Ypsilantis' proclamation, petitions sent by the Muslims of the Morea, and official reports convinced the Porte that the real aim of the insurgents was to annihilate the Muslims of the province at all costs.[82]

In his account, Mir Yusuf, a Muslim Moreote, acutely observed the ideological shift in the 1820s that was epitomized in the fall of Tripoliçe: "The Greek nation", not only slayed all the Muslims, but also turned their cemeteries into fields of barley. For him, this had no precedence in world history. In the negotiations for the capitulation of Anabolu, the Greek delegation explained to the Ottoman side that the oppression of the pashas was the sole reason for the revolt, whose ultimate goal was the foundation of a European-style independent government with its own laws. But Yusuf believed that what accounted for the mayhem was the religious hatred and eternal enmity that the Greeks had bred since the time of the conquest four centuries ago. In a revealing conversation, Yusuf was astonished by the determination of John Orlando of Hydra when the latter swore that should the revolt fail, all insurgents would seek asylum in Europe rather than submit to the Sublime Porte.[83]

A Turkish mounted irregular from Tokat, Deli Mustafa (Kabudi Mustafa Vasfi Efendi) who fought against the Greek insurgents in the Morea, narrated the protracted melee in the province in the usual *gazi*-warrior ethos without a single mention of the revolutionary principles.[84] However, a close reading of his memoirs reveals that he hardly differed from Sultan Mahmud II, or Mir Yusuf in his understanding of the problem. The insurgents killed the Muslims indiscriminately and thus invited the Muslim retaliation. For him, state-sanctioned violence exercised on the Greeks in insurgent regions was permissible in order to force them back to Ottoman subjecthood. He defined the Greeks as *kafir* (infidel) or *asi* (rebel) as opposed to *re'aya* (non-Muslim subjects), and reserved ethno-linguistic terms for the Muslims: Turks, Kurds, and Albanians. In close resemblance to Mahmud II, he condemned the Albanians for their lack of religious zeal in the fight against the insurgents. He viewed them as rebellious and selfish brigands, who did not know any Turkish. He vaguely linked "political loyalty and Muslimness to knowing Turkish." On one occasion, he was not easily convinced of the Muslimness of a group of village women, as they could not speak Turkish.[85]

Particularly, the news of extermination of the Muslim population of Triboliçe that occurred after the Albanian garrison capitulated the town to Greek revolutionaries, was shocking to Istanbul. Unsure of the religious zeal and political loyalty of the native Albanians of the Morea, the Porte decided to rely on "Turkish lads" to suppress the revolt in the Morea. While this novelty did not necessarily mean an ideological shift that favored Turkish nationalism, it was likely to enforce ethno-religious boundaries; it, thereby, fits very well the Mediterranean and local realities.[86]

Interestingly, the Spanish liberals also inspired Mahmud II in whose imagination they "defied the entire world with the strength they derived from solidarity" in a country as small as the palm of a hand. Mahmud contrasted them with the Muslims, "who idly jiggled their rosaries in the mosques" instead of fighting the infidel.[87]

As late as the 1840s, rumors concerning the Morea abounded in the streets of Istanbul. Some joyfully speculated about a possible Ottoman or Egyptian expedition to the Morea and made loud promises of revenge against the Greeks by joining any campaign on the Morea. Others dreamt about establishing claims to their lost estates in Tripoliçe. Men of all denominations shared the hatred for the extravagant King Otto—"the Bavarian Jew"—and were dismayed at the constant interference of the Great Powers with "the subjects of the Morea" (*tebaa-i Mora*). After the flight of the King, expectations in Istanbul streets ran high that the Great Powers would return the island to the Ottomans, a rumor supported by the Moreote merchants in the city as well.[88]

What happened to the "republic of *ayan*, *hakim*, and *kocabaşı*"? Political independence did not mean a sudden detachment of the province from the Ottoman world. The new state required a different power configuration, and the Adriatic ceased to be a frontier zone. Accordingly, the change of the local context was gradual but definite. The Mediterranean context, on the other hand, seems to have been dominated by Mediterraneanism/meridionalism,[89] a sub-category of Orientalism, which presents the Morea as the heartland of "lazy Greeks" as a discourse found in the recent debate over Greece's debt crisis.

Notes

1. Seyit Ali Kahraman-Yücel Dağlı-Robert Dankoff eds., *Evliya Çelebi Seyahatnamesi* vol. 8 (İstanbul: Yapı Kredi Yayınları, 2003), 128; This was likely a fictional story based on a local hearsay perhaps since Romans had dreamt of a canal through the Isthmus of Corinth several times. Emperor Nero had actually begun a canal construction project at the Isthmus in AD 67; the Venetians also entertained this idea, after they conquered the Morea in 1687. See M. J. T. Lewis, "Railways in the Greek and Roman World," in eds. E. Guy and J. Rees, *Early Railways. A Selection of Papers from the First International Early Railways Conference,* (London: Newcommen Society, 2001): 8–19, 9; Peter Lock, *Franks in the Aegean, 1204 1500* (New York: Routledge, 2013), 68–132.

2. Christy Constantakopoulou, *The Dance of the Islands: Insularity Networks, the Athenian Empire, and the Aegean World* (Oxford and New York: Oxford University Press, 2007), 173. F. Braudel, *The Mediterranean and the Mediterranean World in the Age of Philip II* (Los Angeles, University of California Press, 1972), 160–161.

3. See relevant entries "el-ceziret," "el-cezr," "el-cezer," in Mütercim Asım Efendi, *El-Okyanusu'l-Basit Fi Tercemeti'l-Kamusi'l-Muhit (Kamusu'l-Muhit Tercümesi)*, vol. 2 (Istanbul: Türkiye Yazma Eserler Kurumu Başkanlığı, 2013), 1820, 1818, 1819. The Turkish *ada/atag* means "a separated territory," Andreas Tietze, *Tarihi ve Etimolojik Türkiye Türkçesi Lügatı*, vol. 1 (Istanbul-Wien: Simurg, 2002), 95.

4. Mehmed Neşri, *Kitab-ı Cihan-nüma Neşri Tarihi*, vol. 2, ed. Faik Reşit Unat and Mehmed A. Köymen (Ankara: Türk Tarih Kurumu, 1995), 631.

5. *Evliya Çelebi Seyahatnamesi*, 134. *Al-Jazirah*s in Upper Mesopotamia and Sudan are also imaginary islands.

6. Constantakopoulou, *The Dance of the Islands*, 11. Peregrine Horden and Nicholas Purcell, *The Corrupting Sea: A Study of Mediterranean History* (Oxford: Blackwell, 2000), 77, 382, 392.
7. Braudel uses the term "miniature continents," F. Braudel, *The Mediterranean*, 150–151; Cyprian Broodbank, *An Island Archaeology of the Early Cyclades* (Cambridge: Cambridge University Press, 2002), 6–27.
8. Antonis Hadjikyriacou, "Society and Economy on an Ottoman Island: Cyprus in the Eighteenth Century," (Ph.D. diss., SOAS, University London, 2011).
9. Clifford Geertz, "Thick Description: Toward an Interpretative Theory of Culture," *The Interpretation of Cultures: Selected Essays* (New York: Basic Books, 1973), 3–30.
10. Broodbank, *An Island Archaeology*, 16–17.
11. Richard Stites, *The Four Horsemen. Riding to Liberty in Post-Napoleonic Europe* (New York: Oxford University Press, 2014), 3.
12. Ibid., 6.
13. For a comparative analysis of the Epidaurus Provisional Constitution of 1822, see ibid., 214–22. For international revolutionaries in the Morea, see ibid., 232–237.
14. Ibid., 323.
15. Kahraman Şakul, "Adriyatik'de Yakobinler: Mehmed Şakir Efendi'nin 'takrir-gune' Tahriri," *Kebikeç* 33 (2012): 231–251. On Ali Pasha, see Frederick Anscombe, "Continuities in Ottoman Centre-Periphery Relations, 1787–1913," in *Frontiers of the Ottoman World*, ed. A.C.S. Peacock (Oxford University Press, 2009), 235–253.
16. Başbakanlık Osmanlı Arşivi [BOA] Cevdet Hariciye Kataloğu [C. HRC] 9196 (April 12, 1804).
17. BOA, Hatt-ı Hümayun Kataloğu [HAT] 176/7672 (July 3, 1799); İsmail Hakkı Uzunçarşılı, "Arşiv Vesikalarına Göre Yedi Ada Cümhuriyeti," *Belleten I*/34 (1937), 636–37.
18. Daniel Panzac, "International and Domestic Maritime Trade in the Ottoman Empire during the 18th Century," *International Journal of Middle East Studies* 24, no. 2 (1992): 204.
19. Gelina Harlaftis and Sophia Laiou, "Ottoman State Policy in the Mediterranean Trade and Shipping, c. 1780–1820: The Rise of the Greek-Owned Ottoman Merchant Fleet," in *Networks of Power in Modern Greece*, ed. Mark Mazower (London: Hurst, 2008), 1–46.
20. Eloy Martin Corrales, "Greek-Ottoman Captains in the Service of Spanish Commerce in the Late Eighteenth Century," in *Trade and Cultural Exchange*, eds. Maria Fusaro, Colin Heywood, and Mohamed Salah Omri

(New York: I.B.Tauris, 2010), 203–223; G. Harlaftis, "The Eastern Invasion: Greeks in Mediterranean Trade and Shipping in the Eighteenth and Early Nineteenth Centuries," in ibid., 223–252.

21. John Cam Hobhouse, *A Journey through Albania and other Provinces of Turkey in Europe and Asia, to Constantinople, During the Years 1809 and 1810*, vol. 1 (Philadelphia: M. Carey and son, 1817), 186–187; Henry Holland, *Travels in the Ionian Isles, Albania, Thessaly, Macedonia, etc. during the years 1812 and 1813* (London: Longman, Hurst, Rees, Orme, and Brown, 1815), 433; Gerasimos M. Vlachos, *Επτάνησος Πολιτεία: Ο Κανονισμός της Ναυτιλίας (1803/1805)* [*The Merchant Marine Regulation of the Ionian Republic (1803/1805)*] (Athens: s.l., 2005).

22. Norman Saul, *Russia and the Mediterranean 1797–1807* (Chicago: University of Chicago Press, 1970); Kahraman Şakul, "Ottoman Attempts to Control the Adriatic Frontier in the Napoleonic Wars," in *The Frontiers of the Ottoman World*, ed. A.C.S. Peacock, 253–271.

23. Panagiotis Stathis, "From Klepths and Armatoloi to Revolutionaries," in *Ottoman Rule and the Balkans, 1760–1850: Conflict, Transformation, Adaptation*, eds. Antonis Anastasopoulos and Elias Kolovos (Rethymno: University of Crete, Department of History and Archaeology, 2007), 167–179.

24. BOA, C.HRC 367 (September 1798), C.HRC 1491 (December 9, 1798), C.HRC 2320 (November 2, 1798).

25. BOA, HAT 168/7123 (December 29, 1797–January 8, 1798); Fatih Yeşil, "Looking at the French Revolution through Ottoman Eyes: Ebubekir Ratib Efendi's observations," *Bulletin of the School of Oriental and African Studies* 70, no. 2 (2007): 283–306.

26. BOA, HAT 168/7123 (late 1797).

27. BOA, C.HRC 2025 (September 1798). The letter sent to the Porte was in Greek. The Lalliotes were known to be fierce bandits. Like the people of Bardunia, they were also Greek-speaking Muslims.

28. BOA, Bab-ı Asafi Amedi Kalemi Dosyaları [A.AMD] 39/12 (July 26, 1797), HAT 1192/46903-C (July 26, 1797), 171/7331 (July 26, 1797).

29. BOA, HAT 176/7712 (February 13, 1802); HAT 261/15092 (updated); Hüseyin Şükrü Ilıcak, "A Radical Rethinking of Empire: Ottoman State and Society during the Greek War of Independence: 1821–26," (Ph.D. diss., Harvard University, 2011), 197–202.

30. Dean J. Kostantaras, "Christian Elites of the Peloponnese and the Ottoman State, 1715-1821," *European History Quarterly* 43, no. 4 (2013): 637–638, 648–649. There were 20 Muslims and 21 Christians in the plot, see footnote 61 on page 652.

31. Demetrios Stamatopoulos, "Constantinople in the Peloponnese: The Case of the Dragoman of the Morea Georgios Wallerianos and Some Aspects of the Revolutionary Process," in *Ottoman Rule and the Balkans*, eds. Antonis Anastasopoulos and Elias Kolovos, 155.
32. For the list of these *kapudanlık*s, see Fariba Zarinebaf, "Soldiers into Tax-Farmers and Reaya into Sharecroppers: The Ottoman Morea in the Early Modern Period," in Zarinebaf, John Bennet and Jack L. Davis, *A Historical and Economic Geography of Ottoman Greece. The Southwestern Morea in the 18th Century* (Athens: School of Classical Studies, 2005), ft. 89, 28.
33. Ibid., 19, 28–29, 31.
34. Jane Hathaway, "The Household: An Alternative Framework for the Military Society of Eighteenth Century Ottoman Egypt," *Oriente Moderno* 18, no. 1 (1999): 57–66.
35. Ali Yaycıoğlu, "Provincial Power-Holders and the Empire in the Late Ottoman World: Conflict or Partnership?," in *The Ottoman World*, ed. Christine Woodhead (New York: Routledge, 2012), 436–451; Ali Yaycıoğlu, "The Provincial Challenge: Regionalism, Crisis, and Integration in the late Ottoman Empire (1792–1812)," (Ph.D. diss., Harvard University, 2008), 159–160; Demetrios Papastamatiou, "Tax-Farming (*İltizam*) and Collective Fiscal Responsibility (*Maktu*) in the Ottoman Southern Peloponnese in the Second Half of the Eighteenth Century," in *The Ottoman Empire, The Balkans, the Greek Lands: Towards a Social and Economic History. In Honor of John C. Alexander*, eds. Elias Kolovos, Phokion Kotzageorgis, Sophia Laiou, Marinos Sariyannis (Istanbul: Isis, 2007), 289–307.
36. Yaycıoğlu, "Provincial Power-Holders," 446–447.
37. Jane Hathaway, "Rewriting Eighteenth-Century Ottoman History," in *Mediterranean Historical Review* 19, no. 1 (2004), ed. Amy Singer: 29–53; Dina Khoury, "The Ottoman Centre Versus Provincial Power-Holders," 135–156; Fikret Adanır, "Semi-Autonomous Forces in the Balkans and Anatolia," in *The Cambridge History of Turkey*, vol. 3*: The Later Ottoman Empire, 1603–1839*, ed. Suraiya N. Faroqhi (Cambridge: Cambridge University Press, 2006), 157–185; Bruce Masters, "Semi-Autonomous Forces in Arab Provinces", in ibid., 186–208.
38. Kahraman Şakul, "An Ottoman Global Moment: War of Second Coalition in the Levant" (Ph.D. diss., Georgetown University, 2009), 255–310.
39. BOA, HATT 176/7666–B (before August 25, 1803); see Şakul, "An Ottoman Global Moment," 364–365.
40. Papastamatiou, "Tax-farming (*İltizam*)," 292.
41. Antonis Anastasopoulos, "The Mixed Elite of a Balkan Town: Karaferye in the Second Half of the Eighteenth Century," in *Provincial Elites in the Ottoman Empire: Halcyon Days in Crete V. A Symposium Held in*

Rethymnon, 10-12 January 2003, ed. Antonis Anastasopoulos (Rethymnon: Crete University Press, 2003), 259–268.

42. Stamatopoulos, "Constantinople in the Peloponnese," 149–150.
43. "Pek güzel! Lakin cezire-i Mora kaza itibarıyla yirmidir. Onaltısının ashabı malum. Şunda dört kaza cezirenindir. Onaltı kazada rüyet-i maslahat valinin istediği vakitde suret bulmak mertebesinden haric olduğu tarif istemez. İşte şimdi zuhur iden emr-i ali Kartina voyvodası Hacı Hasan Ağa'nın uhdesinde olan mukataat cizyelerinin hesabı rüyetidir. Ve Kartina Kocabaşısı Deli Yani'nin hesabı maddesidir. Mazmun-ı emr-i ali üzere bu gün mübaşeret eylesem 10 güne kalmaz başka suret emir getürürler. Ve mukataa sahibesi tarafından iğbirar derkar olur. Buna çare nedir sultanım? İşe şuru olunmaksızın evvela el-emr agah buyurub münasib görülür ise hesab maddesi bervech-i hakkaniyet rüyetimize ruhsat hükmü alınub gönderilmek kabil midir? Ki tarafeyni gadr ve himayeden ari hakkaniyet vechle hesablarına feysal virilüb hakikatı arz ve inha oluna. Bu maddeye cenab-ı şerifden re'y-i sa'ibane ve tedbir-i sakıbane isterim. İşe mübaşeret eylediğimizde yine başka suret tenbih gelecek ise kerem idüb bizi ol-tarafda lisana getürmeyün. Eğer tarafımıza ve hakkaniyetimize havaleye her taraf razı olur ise azizim ruhsatname almağa muhtacdır. Sözün dost-doğrusu budur. Heman ne vechle iktiza ider ise tizcek cevab irsalinize muntazırım. İfşa ve istişaresi münasib olanlar ile müzakere buyurub neticesini aceleten tahrire himmetleri mercudur." BOA, C.HRC 7283 (August 3, 1801).
44. Zarinebaf, "Soldiers into Tax-Farmers," 39.
45. Stamatopoulos, "Constantinople in the Peloponnese," 155.
46. "The land of Greeks consisted of 5 pachaliks, 2 vaivodalics [sic], and an infinite number of governments entrusted to officers of inferior rank." Tennet James Emerson, *The History of Modern Greece*, vol. 1 (London: H. Colburn and R. Bentley, 1830), 283. Hobhouse translated *kaza* as "canton governed by a Greek codja-bashee, or elder." Hobhouse, *A Journey through Albania*, 192.
47. "[…] her bir kazā ahālisi birer vechile semt-i selāmeti cüst-ı cuyā zāhib olarak Balya Badra kazāsı misillü serbestiyetnāme tahsîline ve ba'zı kazālar selātin-i 'izām hazerātının 'uhdelerinde olmağla anlar dahî istiklāl ve serbestiyyet üzre emirler ısdārına destres olub Mora vālilerinin zulmiyyelerini tahmîl idecek fakat beş altı kazā kalub mā'dāsı Mora'dan müfrez gibi olmağla serbestiyetnāme tahsîliyle mümtaz olan selātin-i 'izām mukāta'atı re'āyaları dahî taraflarından mansub voyvodaların zulm ve sitemîne ibtilā ile el-hālet-i hāzihî Mora'nın bir tarafı vālilerin zulmuyle ve taraf-ı diğeri voyvodaların gadr ve te'addî ve sitemiyle harāb ve vîrān […]." Fatih Yeşil and Yunus Koç, eds., *Nizam- Cedid Kanunları* (1791–1800) (Ankara: Türk Tarih Kurumu, 2012), 43–49.

48. For the 1812 law, see Fatih Tayfur, *Osmanlı Belgeleri Işığında 1821 Rum İsyanı ve Buna Karşı Oluşan Tepkiler* (MA diss., Marmara University, 2003), 76–84.
49. Şakul, "An Ottoman Global Moment," 352–74.
50. Şakul, "Military Transportation as Part of Mediterranean Maritime Trade: Ottoman Freight Payments during the War of the Second Coalition (1798–1802)," *Journal of Mediterranean Studies* 19 (2010): 392.
51. Zarinebaf, "Soldiers into Tax-Farmers," 37–39, 194–195, 213.
52. William Gell, *Narrative of a Journey in the Morea* (London: Longman, Hurst, Rees, Orme, and Brown, 1823), 65–66.
53. Aziz Berker, "Mora İhtilali Tarihçesi veya Pehah Ef. Mecmuası, 1769," *Tarih Vesikaları* 2, no. 10 (1942): 318.
54. *Evliya Çelebi Seyahatnamesi*, vol. 8, 138.
55. Ibid., 157.
56. Ibid., 132, 157.
57. For Evliya's description of the Morea, see ibid., 125–171.
58. Frederick F. Anscombe, "Albanians and 'Mountain Bandits," in *The Ottoman Balkans, 1750–1830*, ed. Frederick F. Anscombe (Princeton: Markus Wiener Publishers, 2006), 92–103; Baki Tezcan, "Ethnicity, Race, Religion, and Social Class: Ottoman Markers of Difference," in *The Ottoman World*, ed. Christine Woodhead, 159–171.
59. Zarinebaf, "Soldiers into Tax-Farmers," 17, 212.
60. Gell, *Narrative of a Journey in the Morea*, 104, 157.
61. Hayati Develi, ed., *XVIII. Yy İstanbul'a Dair Risale-i Garibe* (Istanbul: Kitabevi, 1998), 22, 33, 41.
62. Molly Greene, *A Shared World: Christians and Muslims in the Early Modern Mediterranean* (Princeton: Princeton University Press, 2000), 78–110.
63. Virginia H. Aksan, *Ottoman Wars 1700–1870. An Empire Besieged* (Harlow: Longman/Pearson, 2007), 189–91; Berker, "Penah Ef. Mecmuası," 309–310.
64. Ibid., 311.
65. Selim Deringil, *The Well-Protected Domains: Ideology and the Legitimation of Power in the Ottoman Empire, 1876–1909* (London and New York: I.B. Tauris, 1998), 101–103.
66. Antonis Anastasopoulos, "Albanians in the Eighteenth-Century Ottoman Balkans," in *The Ottoman Empire, the Balkans, the Greek Lands*, eds. Elias Kolovos, Phokion Kotzageorgis, Sophia Laiou, and Marinos Sariyannis, 43–44. Ottoman documents define Albanians as tribe (*kabile*), group (*taife*), eth-

nic group (*cins*), people (*millet*), see Hakan Erdem, "'Perfidious Albanians' and 'Zealous Governors': Ottomans, Albanians, and Turks in the Greek War of Independence," in *Ottoman Rule and the Balkans*, eds. Antonis Anastasopoulos and Elias Kolovos, 213–243.

67. Zarinebaf, "Soldiers into Tax-Farmers," 34, 37.
68. Kostantaras, "Christian Elites of the Peloponnese," 634.
69. Johann Strauss, "Ottoman Rule Experienced and Remembered: Remarks on Some Local Greek Chronicles on the Tourkokratia," in *The Ottomans and the Balkans: A Discussion of Historiography*, eds. Fikret Adanır and Suraiya Faroqhi (Leiden: Brill, 2002), 210.
70. Stamatapoulos, "Constantinople in the Peloponnese."
71. Vahdettin Engin, "Mahmud Raif Efendi Tarafından Kaleme Alınmış İngiltere Seyahati Gözlemleri", in *Prof. Dr. İsmail Aka'ya Armağan* (İzmir 1999): 135-162; Kemal Beydilli and İlhan Şahin, *Mahmud Raif Efendi ve Nizam-ı Cedid'e Dair Eseri* (TTK, Ankara, 2001); M. A. Yalçınkaya, "Mahmud Raif Efendi as the Chief Secretary of Yusuf Agah Efendi, The First Permanent Ottoman-Turkish Ambassador to London (1793–1797)," *Osmanlı Tarihi Araştırma Merkezi* V (1994): 422–434.
72. Maurice Herbette, *Fransa'da İlk Daimi Türk Elçisi "Moralı Esseyyit Ali Efendi" (1797–1802)*, trans. E. Üyepazarcı (Istanbul, 1997).
73. Haşim Koç, "17.yy Ortasında Osmanlı Coğrafyasından Antik Dönemlere Bir Bakış: Katip Çelebi'nin Eserlerinden Seçmeler," *Doğu Batı* 40 (2007): 257–282.
74. Gülçin Tunalı, *Another Kind of Hellenism? Appropriation of Ancient Athens via Greek Channels for the Sake of Good Advice as Reflected in Tarih-i Medinetü'l-Hukema* (Bochum: Ruhh University, 2013), 32, 83.
75. Ibid., 175; Gülçin Tunalı, "*Tarih-i Medinetü'l-Hukema* or the History of Ancient Athens from Intercultural History Perspective?," in *10th Annual International Conference on Conceptual History: Transnational Concepts, Transfers and the Challenge of Peripheries*, ed. Gürcan Koçan (Istanbul: ITU, 2008).
76. Gülçin Tunalı, "Osmanlı Atinası ve Düşünce Tarihi Ekseninde Kadı Mahmud Efendi'nin *Tarih-i* Medinetü'l-Hukema Adlı Eseri," *Divan İlmi Araştırmalar* 20, no. 1 (2006): 169–184; Tunalı, *Another Kind of Hellenism*, 124.
77. The Ottoman conquest of 1715 is narrated in folios between 241a and 291b. The transcription of Mahmud Efendi's account leaves this section out. See Appendix 2 (pages 208–359) for the transliteration of the text in Tunalı, *Another Kind of Hellenism?*, 83.
78. Gell, *Narrative of a Journey in the Morea*, 304.

79. Ilıcak, "A Radical Rethinking of Empire," 197.
80. Ahmed Aydın, "Mir Yusuf Tarihi (Metin ve Tahlil)" (MA diss., Marmara University, 2002), 36–37, 47, 55.
81. Nicholas C. Pappas, *Greeks in Russian Service in the Late Eighteenth and Early Nineteenth Centuries* (Thessalonica: Balkan Institute, 1991), 325–337. At least 6,500 "Greeks" (Orthodox Christians) served in the armies of Russia, France, and the Kingdom of Naples in 1807. See Stathis, "From Klepths and Armatoloi to Revolutionaries."
82. Anscombe, "Albanians and 'Mountain Bandits;'" Ilıcak, "A Radical Rethinking of Empire," 1–13, 67–68, 101, 147.
83. Aydın, "Mir Yusuf Tarihi," 26, 60–61, 69, 70, 72.
84. Jan Schmidt, "The Adventures of an Ottoman Horseman: The Auto-biography of Kabudlı Vasfi Efendi, 1800–1825," in *The Joys of Philology. Studies in Ottoman Literature, History and Orientalism (1500–1923)*, ed. Jan Schmidt (Istanbul: Isis, 2002), 166–286.
85. Tolga U. Esmer, "The Confession of an Ottoman 'Irregular': Self-Representation and Ottoman Interpretive Communities in the Nineteenth Century," *Journal of Ottoman Studies* [*JOS*] 44 (2014): 331. For more on Turkish contempt about non-Turkish Muslims, also see ft. 53 on page 330. On the views of Mahmud II, see Ilıcak, "A Radical Rethinking of Empire," 162–169, 182.
86. Hakan Erdem, "Recruitment for the 'Victorious Soldiers of Muhammad' in the Arab Provinces, 1826–28," in *Histories of the Modern Middle East: New Directions*, eds. Israel Gershoni, Hakan Erdem, and Ursula Woköck (London: Lynne Rienner, 2002), 189–206; Yusuf Hakan Erdem, "Do not Think of the Greeks as Agricultural Labourers': Ottoman Responses to the Greek War of Independence," in *Citizenship and the Nation-State in Greece and Turkey*, eds. Faruk Birtek and Thalia Dragonas (London and New York: Routledge, 2005), 67–84; Veysel Şimşek, "The First 'Little Mehmeds': Conscripts fort he Ottoman Army, 1826–53," *Journal of Ottoman Studies* 44 (2014): 265–311.
87. Ilıcak, "A Radical Rethinking of Empire," 154.
88. Cengiz Kırlı, *Sultan ve Kamuoyu. Osmanlı Modernleşme Sürecinde Havadis Jurnalleri (1840–1844)* (Istanbul: Türkiye İş Bankası Kültür Yay., 2009), 122, 211, 244, 247, 253, 265, 288, 331, 378, 409, 414, 416, 418, 420, 421, 426, 429, 441.
89. Stites, *The Four Horsemen*, 5.

Contibutors

Spyros I. Asdrachas has taught history in France (Université Paris I, Pantheon-Sorbonne) and organized and conducted research in Greece (National Hellenic Research Foundation). He is one of the most eminent representatives of the *Annales* School in Greece, employing its research tools and agenda for the purposes of reconceptualising Modern Greek history. His two main areas of interest concern the economic history of the Greek populations of the Ottoman Empire, and the islands of the Ionian and Aegean Sea.

Antonis Hadjikyriacou is Assistant Professor of Early Modern Ottoman and Mediterranean History at Boğaziçi University, Turkey. He has been Mary Seeger-O' Boyle Postdoctoral Research Fellow at Princeton University and Marie Curie Intra-European Fellow at the Institute for Mediterranean Studies, Foundation for Research and Technology—Hellas. His teaching and research interests include Ottoman social and economic history, environmental and climate history, spatial history, and historical GIS.

Murat Cem Mengüç is the Director of the Middle Eastern Studies Program and an Assistant Professor of History at Seton Hall University, New Jersey, USA. His publications include "Neşri's Cihannüma: An early Ottoman history book and politics of Ottoman identity," in *Living in the Ottoman Realm*, ed. Christine Verhaaren and Kent Schull (Bloomington: Indiana University Press, 2015); "The Türk in Aşıkpaşazade: An Ottoman history by a private individual," *Journal of Ottoman Studies* 44 (2014); "Histories of Bayezid I, historians of Bayezid II: Rethinking late fifteenth-century Ottoman historiography," *Bulletin of the School of Oriental and African Studies* 76 :3 (2013).

Kahraman Şakul is Assistant Professor at the Department of History, İstanbul Şehir University. He holds BA and MA degrees from the Department of History, Boğaziçi University, İstanbul (1994–2001). He received his Ph.D. degree from the History Department of Georgetown University (2009) for his dissertation entitled "An Ottoman Global Moment: War of Second Coalition in the Levant, 1798–1807." He has published several articles in Turkish and English on Ottoman diplomacy, military and technology, as well as the transformation of the Ottoman political culture in 1774–1826.

Fatma Şimşek is Assistant Professor at the Department of History, Akdeniz University, Turkey. She teaches undergraduate and postgraduate courses on Ottoman paleography, Ottoman institutional history, and modern European history. Her research interests concern Ottoman administrative organization, the social and administrative organization of the Ottoman islands in the Mediterranean, and French trade in the Mediterranean during the eighteenth century.

Michael Talbot is currently a Lecturer in History at the University of Greenwich. He completed his PhD in Ottoman history at SOAS, University of London in 2013, and now researches a variety of topics on Ottoman diplomatic, commercial, and global history from the seventeenth to the twentieth century. His first book, *British-Ottoman Relations, 1661–1807: Commerce and Diplomatic Practice in Eighteenth-Century Istanbul*, was published in 2017. He is a longtime contributor the "Ottoman History Podcast" and its primary source blog, "Tozsuz Evrak".

Eleftheria Zei studied at the Faculty of Letters of the University of Athens and has continued her graduate education at the University of Paris I (Panthéon-Sorbonne), where she obtained her doctorate thesis on the island of Paros and the Aegean between Latin and Ottoman domination. She has been teaching Modern History at the Greek Open University since 2003, and in 2008 she joined the teaching staff at the Department of History and Archaeology of the University of Crete. Her research focuses on the modern history and historiography of insular societies in the Mediterranean. Her publications explore questions of social history of the Greek Archipelago and the island of Crete from the seventeenth to the nineteenth century.

www.ingramcontent.com/pod-product-compliance
Lightning Source LLC
Chambersburg PA
CBHW030141170426
43199CB00008B/155